Motorbooks International

POWERPRO SERIES

HOW TO BUILD CHEVROLET SMALL-BLOCK V-8 RACE ENGINES

Bill Tarrant and Chris Hawkinson

Dedication

This book would not have been written if, way back in 1970,
I had not had the door opened to a world of words and ideas that I never knew existed.
Thank you, Ellen Schrader, a great freshman composition instructor, wherever you may be. My education would never have happened if it were not for the foresight and planning of my parents.
Very special thanks to my wonderful wife, Jenny,
for encouraging me and sticking by my side throughout it all.
—Bill Tarrant

First published in 1993 by Motorbooks International Publishers & Wholesalers, PO Box 2, 729 Prospect Avenue, Osceola, WI 54020 USA

Library of Congress Cataloging-in-Publication Data

Tarrant, Bill.
How to build Chevrolet small-block V-8 race engines / Bill Tarrant, Chris Hawkinson.
p. cm. — (Motorbooks International powerpro series)
Includes index.
ISBN 0-87938-783-1
1. Automobiles, Racing—Motors—Design and construction. 2. Chevrolet automobile—Motors—Modification. I. Hawkinson, Chris. II. Title. III. Series.
TL210.T37 1993
629.25'04—dc20 93-26340

On the front cover: A race-ready Chevrolet 350ci small-block that has been built with an attention to detail but without exotic and expensive aftermarket equipment that's beyond the budget of most grass-roots race teams. *Bill Tarrant*

Printed and bound in the United States of America

Contents

Preface

How to Build a Chevrolet Small-Block Race Engine presents a hands-on approach to building a small-block engine for the car enthusiast on a budget. That category includes most of us who work for a living. Therefore, we will not discuss exotic machine work and expensive parts. And we will not even *mention* billet cranks, aluminum blocks, changing of valve angles in the heads, or anything of that nature since, although these goodies have their place, they are not for the backyard enthusiast or hobby racer.

Chevy enthusiasts of all types can benefit from reading this book. The stock car racer from street stock to pro-stock and modified, classes that run under a claim rule, can learn to build an economical engine. The drag racer who races in a bracket class can learn to build reliable, consistent horsepower. And the street enthusiast can learn to build a fun-to-drive, dependable street engine.

The information presented in the book comes from firsthand experience and is not based solely on theory. Co-author Christopher "Wheat" Hawkinson is an experienced engine builder, building late-model engines for ten years before joining the International Motor Contest Association (IMCA) ranks eight years ago. Wheat did not have the luxury of unlimited funds in his late-model racing, so he learned by trial and error what he could get by with and what he couldn't, which stock parts would work and which ones wouldn't, and which aftermarket pieces were necessary. This background proved to be vital for Wheat's success, one that he applies to his IMCA-claim engines.

Acknowledgments

We wish to thank Wayne Palmer, our employer at Galesburg Machine Shop in Galesburg, Illinois, for allowing us to use his shop to build the engines. We also wish to thank him for providing many of the parts that went into these engines. Without him, this project would have been difficult, if not impossible, to accomplish.

Special thanks also goes to Larry Wright, auto mechanics instructor at Carl Sandburg Junior College in Galesburg, Illinois. Larry tested the engines on the college dynamometer so we could provide readers with horsepower figures for these do-it-yourself engines.

Thanks to Verne Schumann of Schumann's Sales and Service, Blue Grass, Iowa, for help procuring the Dart S/R heads.

Finally, a thank you goes to the fine people at Midwest Photo and Video in Galesburg for the film processing and picture printing. They had a lot of patience with us.

How to Build Claimer Race Engines 1

Building a fast car, whether for the drag strip, circle track, or the street, is an American institution, an institution that brings together spouses, brothers, sisters, and cousins, as well as friends, to join in the labor of expression through building an individualized piece of machinery. This segment of Americana has been in danger of dying, not through lack of interest, but because it is becoming too expensive. Those who race as a hobby in the bracket wars on the drag strip or on Saturday night dirt stock car tracks, do not have major sponsors (sponsors at this level of the sport do not contribute thousands of dollars, but $50 and $100). With the escalating cost of tires, fuel, and parts—as well as the everyday cost of living—it has become increasingly difficult for the hobby racer to remain competitive. Many veterans gave it up a few years ago, and aspiring racers saw racing as only an impossible dream. Bracket racing has given drag racers a break, one where consistency is more important than speed. Stock car racers have been blessed with engine claim racing to help equalize competition.

As stated earlier, the information contained in this book is geared toward such enthusiasts— those who work on a modest budget and do most of their own work. The more you understand the basics of engine building, the hows and whys, and the more you can do yourself, the more you will save. Plus, there is a certain amount of personal satisfaction in stepping back, looking at the completed, running product sitting there in the chassis, and being able to say, "I did that myself." We hope to convey to you knowledge gained over the years so that you do not have to go through so much trial-and-error. The principles presented here are tried and true, and are incorporated into everyday engine building. There are no "tricks," so to speak.

Project Engines

We will present two Chevrolet engines: a 0.030in over 350ci and a 0.030in over 400ci.

The 350ci (or 355ci with the overbore) is a very streetable engine, one that you can put in your vehicle and drive across the country. Still, it puts out more than 100hp over stock and produces excellent torque in a good, flat curve. This is a fun engine to drive and requires no more maintenance than a factory stock engine.

The 400ci (406ci with the 0.030in overbore), on the other hand, is not streetable, at least not in practical terms. It is intended for the racetrack only, utilizing a solid-lifter cam, extra-capacity oil pan, and 12.0:1 compression.

Both engines utilize stock and nearly stock pieces—there are no exotic rods or crankshafts, nor is there any elaborate machine work. These are cast-iron engines, including the heads, that can be built without breaking the bank. In fact, if you have a 350ci engine in your car now that is in good condition and not using oil, you can install the upper end, that is, heads, manifold, and cam, that we used on

The inside of Dick Walsh's garage. Dick is an IMCA modified racer from Galesburg, Illinois. He races under an engine claim rule, therefore he tries to keep engine costs to a minimum. Racing is not Dick's profession, but his hobby. His workshop is professional, though. Note the cleanliness, and the air compressor.

Dick Walsh has a neat, clean workbench. All tools are put away; nothing is left lying around to get lost or damaged. All space is utilized as well, even the area beneath the bench, but it is done in an orderly fashion. You can bet that every tool in that toolbox is in its place, too.

our 355 and achieve the same results.

Building Your Race Engine

So, what does it take to learn to build an economical Chevrolet racing engine, whether it be for the circle track, the drag strip, or the street? Actually, it is a combination of several elements, all of which overlap. Knowledge, skill, patience, attitude, and working conditions and equipment are all vital elements. If you only possess one or two of these ingredients, you will likely be unsuccessful, spending too much money or building engines that won't last or make enough power. If you have the most well-equipped, spacious garage in the country, but don't have the knowledge or skill to utilize these conditions, you will not succeed. If you have the skill, but lack the most basic facilities and tools, you won't have the means to apply your skill. So, it is a combination of elements that make the successful engine builder.

Attitude may be the most important element of all—it has a bearing on all phases of your operation and will influence the final outcome of your labor. Attitude is the overall way in which you approach the business of engine building. Do you plan to throw together an engine quickly and hope for the best? Or do you want to take as much time as is feasible and check and recheck for correct clearances and assembly? A wrench or shop towel left inside the oil pan or under the intake manifold will wreak havoc quickly with your new engine. Impossible to do such a thing, you say? No, it's not; it has happened. Disasters such as this are perfectly avoidable, too, with a little patience and care.

Work Area and Organization

Attitude is reflected in your work area. A neat, clean, well-organized, well-lighted shop is vital. You don't need a large, professional shop to achieve professional results. In fact, at this level of engine building, such a place is not practical. Work with the facilities you have, whether it be a single-stall garage or something larger. If you don't own such a place, use the basement, or team up with a buddy who has a garage.

Organization in your workshop is vital. It may seem easier and quicker to simply drop tools where they are used, but next time you want them you will have to look through the mess on your cluttered workbench or in the pile of wrenches on the floor. Buy a toolbox and use it. Clean the tools off every time you put them away, and put them away every time you are done using them.

Keep bench tops clean and free of clutter. Keep the floor clean. You may not have a fancy, painted floor, but take care of what you do have. Anything except a dirt floor will be fine. You absolutely *cannot* assemble an engine in dirt! No matter how careful you think you can be, those dirt particles will find their way into the engine. And dirt ruins engines!

As mentioned, it is important to have a well-lighted work area. You will need to see as well as possible to do a proper assembly job. Details won't show up in dim conditions. Hang as many lights as it takes to brilliantly light up your shop.

Organize your spare parts. Lots of shelves and cupboards are invaluable for storing things in an orderly manner. As you continue in your engine building and car racing, you will probably accumulate parts for future use. By keeping these items organized, they will be easy to find when the time comes and they will also be protected from damage. Bins for bolts and nuts are a good idea. Nothing is more frustrating than digging through a box of various size bolts with washers, nuts, and unknown bits of junk mixed in.

Tools

A set of basic hand tools is necessary. Along with this, you will need some specialized tools for engine assembly. One of the most important of these tools is a set of dial calipers. Micrometers are nice and if you can afford them or can get a good deal on them, do so. You can do a lot more with calipers, though, and the investment is much less than for a set of micrometers. You will find yourself using calipers to measure most engine parts from bearings to bolts. They really are invaluable.

You also must have a set of ring expanding pliers, a ring compressor, and a torque wrench. It is impossible to build a dependable engine without these tools.

Just as important as the tools is safety equipment. Always wear safety glasses when grinding, drilling, or hammering. Make sure safety guards are properly in place on power tools such as a bench grinder. Wear a mask over your mouth and nose if you are doing any grinding or wire brushing dirty parts. An injury will take the fun out of things in a hurry. The bottom line is, use safety equipment and

Vertical storage saves valuable floor space, especially in a limited home shop. Certain items are assigned to each shelf so that order is maintained. Notice the rack near the end for hanging spare crankshafts; cranks will, over a period of time, bend if left on the floor. The cranks are bagged to seal out dirt and moisture, as should be any machined engine part. You may not have the luxury of a large garage like this, or this many tools and parts, but the important point is the cleanliness, neatness, and orderliness of it all. Follow this example with whatever you have and your engine building will go smoother.

don't be in a rush and get reckless.

Engine-Building Knowledge and Skill

Knowledge and skill are two elements that take time to learn. Knowledge can be acquired through asking questions of knowledgeable persons, by working with professionals in the field, through reading how-to and other books on the subject, and by trial-and-error in your own shop. Trial-and-error may be one of the best teachers, and, if you pay attention when you make a mistake you will not make it again. Skill comes with trial and error—there is no real way it can be taught, it only comes with practice.

Patience, although difficult for some to maintain, is a necessary ingredient for successful engine building. Patience is part of attitude and will be reflected in your work, from the overall condition of your shop to the smallest detail on your engine. You will need patience when parts do not fit together properly, or when you are not sure of something, or when details and clearances need to be checked. If you feel impatience creeping in, lock the door to the shop, go in the house, and watch television, read a book, take a nap, do something to take your mind off your engine work until you are ready to start fresh. Otherwise, your sloppiness will cost you down the road—or on the track.

With these thoughts in mind, roll up your sleeves, prepare to get dirty, and have some fun building a winning engine.

For serious engine work, these tools are a must: ring pliers for safely installing rings on pistons; ring compressor for putting the piston assembly into the cylinders; and rod bolt boots for protecting the crank. A set of dial calipers will be used often; it is difficult—in fact, nearly impossible—to be a serious engine builder without them. A set of basic hand tools is also a necessity.

Engine Teardown

2

Differences Between the 350 and 400

There are internal dimensional differences between the 350 and 400ci small-block Chevrolet engines and it is important to recognize them. Externally, the engine measurements are the same, so one may be swapped for the other in a chassis with no problem.

Internally, however, differences do exist and amount to more than simply a different bore and stroke. Some of the components vary enough that parts cannot be interchanged without some special machine work. Stroker engines can be built and in fact are quite common, but if you are on a budget, stock bore-and-stroke combinations should be adhered to—special bearings and custom crankshaft grinding can be expensive.

The bore dimensions differ between the two engines. A stock 350ci bore measures 4.000in; a 400 measures 4.125in. If you have a 350 block and want a 400ci engine, how about boring it 0.125in? Impossible! The cylinder walls are barely safe for a 0.060in overbore. If you want a 400ci engine, you will have to get a 400 block. This means, of course, that the pistons are different sizes.

The rods, too, differ in length on the two engines. Rod lengths are measured from the center of the big-end bore to the center of the piston pin bore; this is called center-to-center distance. A 350 rod is 5.700in center-to-center; a 400 rod is 5.560in. The diameter of the crankshaft journal end and the piston pin end is the same between the two rods, though, so it is possible to bolt them to either crank or fasten them to either piston. In the wrong combinations, this will lead to major problems. Either the pistons will pop out of the top of the block, making it impossible to bolt on the heads, or they will not come up high enough in the cylinders, resulting in extremely low compression.

Special pistons are made for racing applications that will allow you to use a 350 rod in a 400. These pistons are common, being manufactured by most piston companies that handle race applications. The reason for using the 350 rod is for optimum rod-to-stroke ratio. A discussion of this is beyond the scope of this book, but it seems that certain rod lengths work better than others, depending upon the stroke of the engine and its intended use.

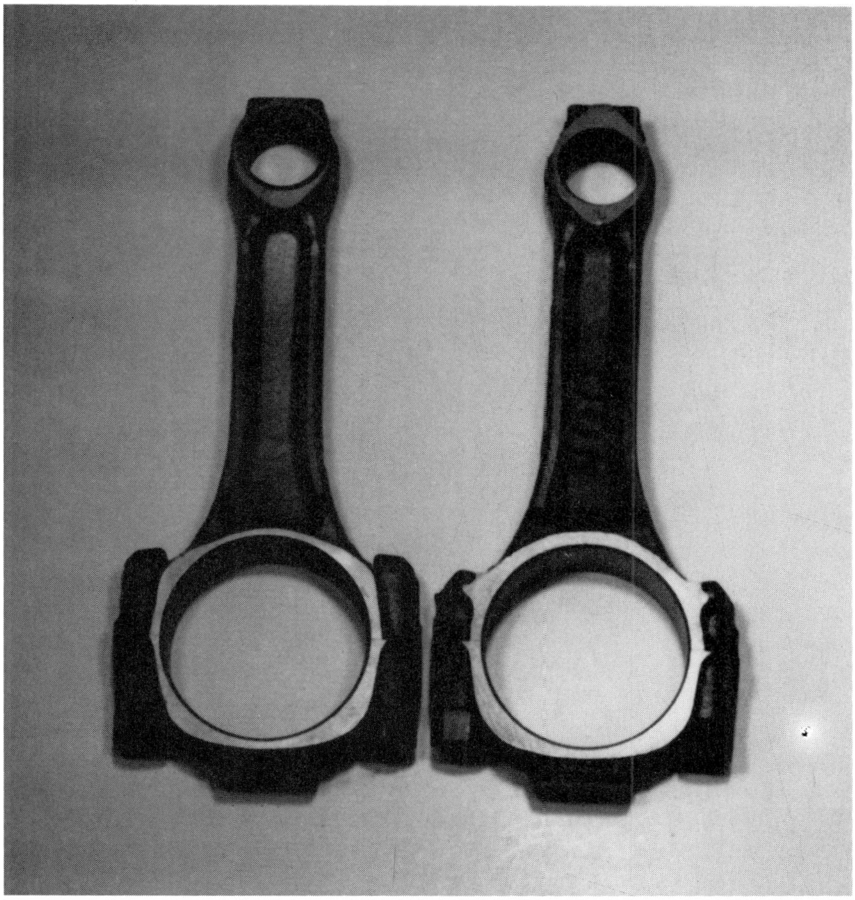

A 350 rod, left, and a 400 rod, right. The 350 rod is 5.700in center-to-center; the 400 rod is 5.560in. This measurement is taken from the center of the rod end bore to the center of the big-end bore, thus the term center-to-center. The longer 350 rod will be used in our 406ci engine, mainly because of piston availability—most performance pistons for the 400ci block require the longer 350 rod.

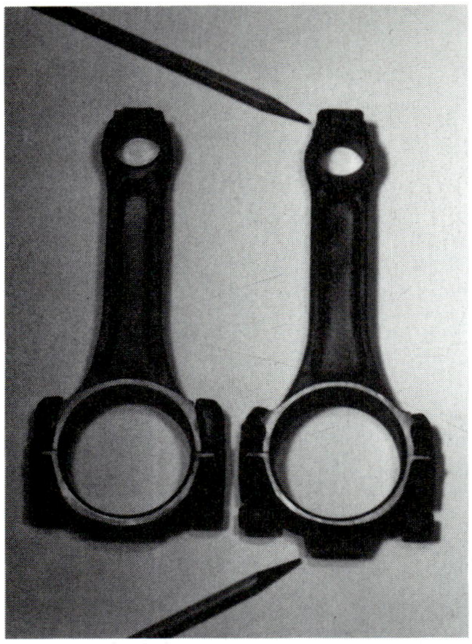

The rod on the right is stronger. You can determine this by the balance pads, those areas of extra casting that the pencils are pointing at. All rods are cast with these pads; the rods are weighed and the pads are trimmed away until the proper weight is reached. When more of the pad has to be cut it means that the bulk of the weight is in the beam of the rod, where it will add strength. Therefore, a rod with the pads trimmed excessively, as the one on the left, should be stronger. Look for rods like this.

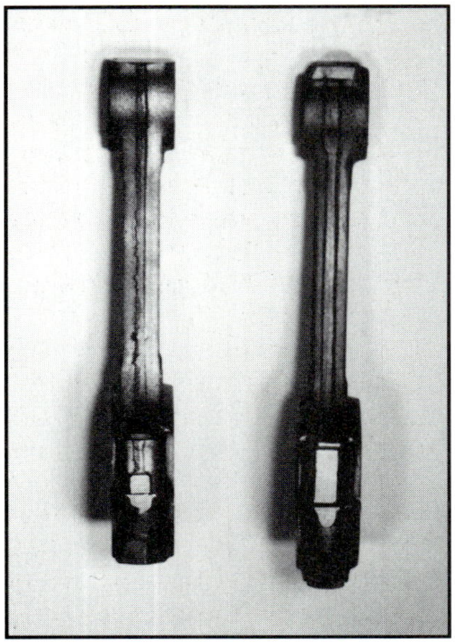

Notice the difference in beam widths. The rod on the left is much thicker, meaning it is stronger. Look for these kinds of rods—they are found in earlier engines and will hold up better.

We had hoped to use the 400 rods in our 406 engine, though, for two reasons. First, if you use a 350 rod you will have to grind the rod bolt heads for clearance between both the sides of the block and the camshaft. That amounts to extra work, and you will be gambling with clearance. Also, the value of power gained with the 5.700in rod over the 5.560in rod in a relatively low horsepower engine is questionable. These are not Winston Cup or Pro-Stock engines here. But, as it turns out, pistons for the stock 400 rod are expensive, so, to keep the cost down we will use the 350 rods.

The crankshaft with the rolled-up piece of paper through the front rod journal is the 400. This is one way to distinguish between the two cranks, though it is not a sure-fire method. Not all 400 cranks have the hollow front journal; a 350 *never* does.

The final check to determine the size of your crank. The 350, left, has no dowel pin in the flange. The 400 crank on the right has a dowel pin to locate the flywheel or flexplate in the correct position. Because it is externally balanced, the flywheel on the 400 has to be positioned a certain way.

The main journals on the two crankshafts are different sizes; the journal size on the 350 crank is 2.4484in to 2.4493in; the journal size on the 400 crank is 2.6488in, plus or minus 0.0005in. In other words, there is at minimum a 0.200in difference in journal sizes. When you are dealing in terms of thousandths of an inch, 0.200in is significant. The crankshafts will not interchange without custom grinding or special bearing inserts. This type of custom engine building is beyond the scope of this book as well, due to both the expense and the custom fitting required to accomplish these special stroker engines. The crankshaft main bores in the blocks, of course, also differ by 0.200in.

To determine which crankshaft you currently have, there are two checks you can make. First, if the first rod journal is hollow it will be a 400ci crankshaft. Note that not all 400 cranks have a hollow journal, but 350 cranks *never* have a hollow journal. So, if it's not hollow it still could be a 400.

The final test is to look at the flange where the flywheel or flexplate attaches. If there is a dowel pin, it is a 400. The dowel is there so that the flywheel or flexplate can be put on in only one position. Incorrectly installed, it will cause a vibration. A dowel is not necessary for the 350, since the flywheel or flexplate may be installed in any position. This leads to one more important difference between the two engines.

A 350 is an internally balanced engine, meaning that balance of the rotating parts is achieved by the parts themselves; that is, the crankshaft has holes drilled in the counterweights to accomplish balance. The 400ci engine, unlike all of the other small-block Chevys, is externally balanced. This means that the balance of the internal parts is accomplished by weighting one side of the harmonic balancer and flexplate or flywheel. A 400 balancer and flywheel, therefore, cannot be used on a 350, or vice versa. To do so will produce a severely vibrating engine.

The flexplate or flywheel for the 400ci engine has a weight added to it, a weight somewhat like a wheel balance weight for balancing tires. The 350 flexplate or flywheel does not have this added weight. The harmonic balancer for the 400 has a cutaway section around the outer circumference to provide balance. The 350 harmonic balancer is perfectly round.

As you can see from these differences, you do not want to buy

The 350 balancer on the right is perfectly round on its outside edge. The 400 balancer has a cutout portion along its outer edge. This is for balance, since the 400 engine is externally balanced. You cannot mix these balancers!

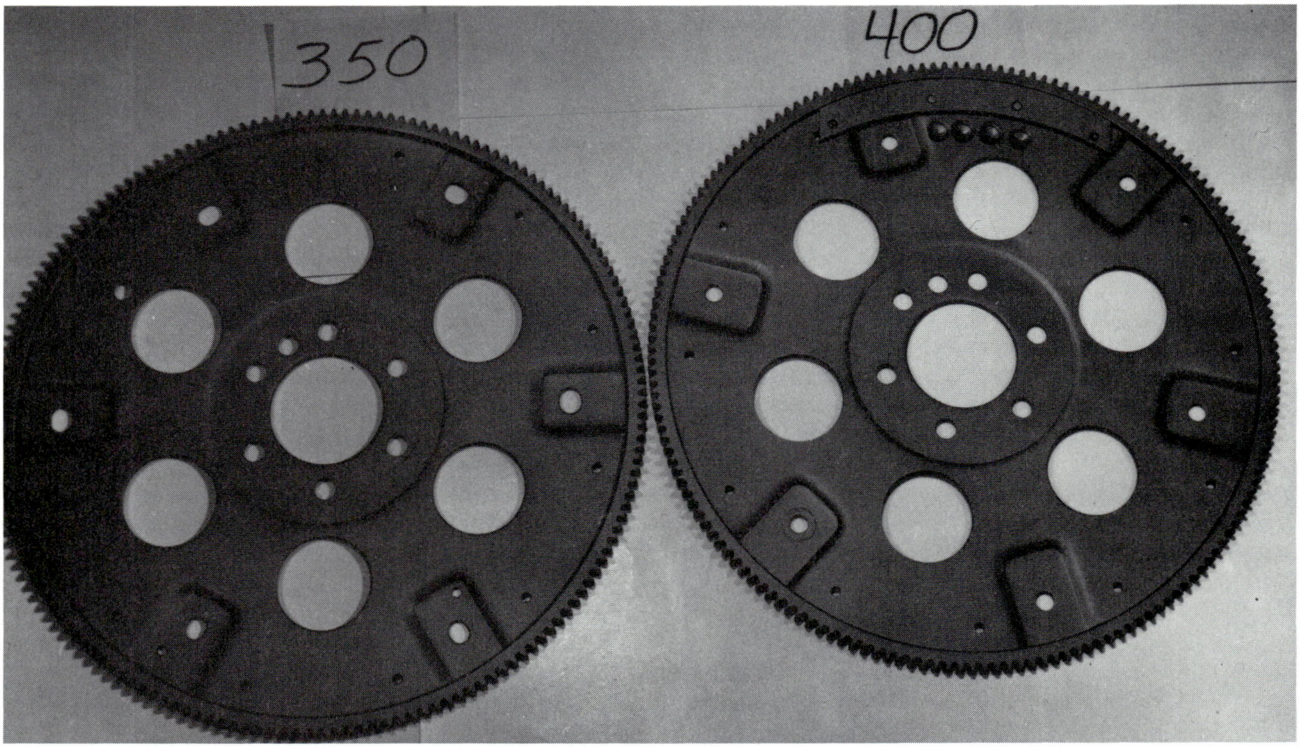

Just as the balancers are different between the two engines, so are the flexplates. The plate for the 350 is unbalanced; the 400 flexplate has balance weights, much like tire balance weights. Like the balancer, the flexplate is part of the external balance for this engine. Do not mix these parts, either.

some parts for a 350 and some for a 400. You would end up with an incompatible pile of engine pieces. The best way to avoid this dilemma is to buy a complete engine. That way, you will have parts that match and you will not have to scrounge for parts that, in the end, don't fit anyway.

Another reason for buying a complete engine is that you get all the small pieces that may not seem important, such as valve covers, valve cover bolts, oil pan bolts, and the like. The cost of these parts adds up in a hurry when you go to buy them separately. Better to begin with them in hand.

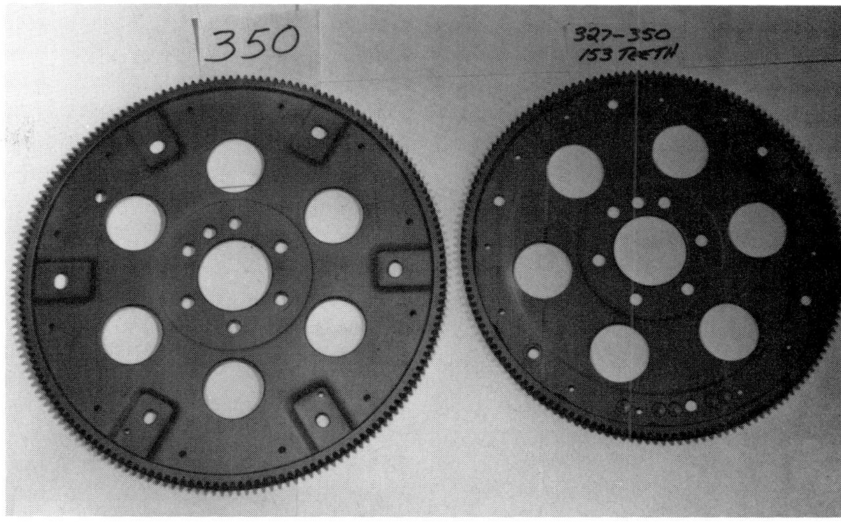

You have a choice of flexplate size on the 350. The smaller one offers less rotating mass, which translates to quicker acceleration. A special starter is required to engage the teeth.

Choosing an Engine to Rebuild

You don't really know if an engine is good unless you hear it run, so if at all possible, listen to the engine run in the car before buying it. This way you can detect any knocks or rattles that will signal useless parts, such as a damaged crankshaft or rods.

If you are considering buying an engine that is out of a vehicle, inspect it closely, looking for holes in the pan or block where a rod has exited. An engine like this is absolute junk—pass it by like the plague.

One simple way to check for damaged bearings is to remove the oil filter and look for the telltale specs of metal. It is not necessary to cut open the filter. Simply look in the top of the filter, down into the little holes. When the oil comes from the block it hits this part of the filter first. Therefore, any pieces of bearing material in the oil will be trapped here.

If you can remove the oil pan before buying the engine, do so—a lot can be learned there as well

The casting number, which is on the bellhousing flange on the same side of the block as the oil filter, is the only number identifying cubic inch displacement.

Determining the Right Vs. Left Side of the Engine

Which are the right and left sides of the engine? This question often leads to confusion, but the answer is simple.

Picture yourself sitting in the driver's seat. You are behind the engine (in a front-engine vehicle). The engine is facing the same direction you are, when looking out the front of the car. Thus, your left and right are the same as the engine's left and right. Just remember this when trying to determine which side of the engine you are dealing with—it will help.

about an engine's condition. Again, as inside the oil filter, look for flecks of bearing material.

If a rod is extremely loose you will be able to rattle it up and down on the crankshaft. A rod cap or main bearing cap that has baked-on oil is a sign of a damaged bearing. The oil will bake on to the cap where the wear occurs due to the extreme heat caused by a spun or worn bearing. Often, too, if this much heat is generated, the crankshaft will be damaged beyond practical repair. The heat will actually bend the crank. An engine with any of these conditions should be avoided.

You may be lucky enough to find a car with enough resale value that you can pull the engine and then sell the body for what you paid for the whole works, thereby getting a free engine. Maybe you can swap some parts you don't need for a complete running engine, or trade some work for one. Use your imagination and ingenuity. Anything to cut costs along the way will be to your benefit, and a good place to start is with the initial purchase. Keep in mind, though, that some deals are too good to be true.

Identifying Engines

Make sure that free engine is not a 307 or some other displacement that you do not want to use. Check casting numbers, if necessary, to play it safe. The block casting number can be found on the rear of the block, on top of the flywheel flange on the left side above the oil filter pad. This is the *only* number on the block that positively identifies it according to displacement.

On the right side of the bellhousing flange you will find another number, such as A 3 2. This number tells you when the block was cast. The letter (in this case A) represents the month: A = January, B = February, C = March, and so on. The following number, which may be single or double digit, tells which day of the month the block was cast. The last number designates the year. This is often a single-digit number, but sometimes may appear as a double, such as 76 and 79. If the final digit is single, a little common sense and further

On the other side of the bellhousing flange is the casting date. It will be in the form of a letter denoting the month (A=January, B=February, and so on), followed by the day of the month, and the year. This number, combined with the casting number, will tell you what displacement the engine is and when it was made.

A pre-1968 small-block. The towers in the front and rear of the lifter valley are the clue. The engine will be 327ci or less.

investigating is needed to determine which decade the number refers to. For example, an 8 could be a 1968, 1978, or 1988 block. There are things to look for on the block that will help narrow down the date.

Under the intake manifold, in the lifter valley, there are more clues. Pre-1968 blocks have a tower in the rear of the valley for the distributor, and a tower in the front of the valley for the oil filler tube in the manifold. A block with both of these towers is a small-journal engine probably of 327 or 283ci. For antique car restorations these engines may be desirable, but for the high-performance applications described here, they are worthless.

A block with a tower only in the rear of the lifter valley is a

This is a 1968–1970 engine; it has a tower in the rear of the lifter valley and could be a 327ci or a 350ci.

pre-1970 and a large-journal engine. Could be a 350 or a 327, though, so check the crankshaft forging numbers if there is any doubt. A good machine shop should have a list of forging numbers—ask them for this information.

Blocks from 1986 to the present use a one-piece rear main seal and have four bolts through the middle of the valve covers instead of around the edges. These two features make them easily recognizable. By keeping these distinguishing features in mind, you can determine what year the engine was made even if there is only a single digit.

Engines 1986–Present

More needs to be said concerning the newer-style engines. They are getting old enough now that they will begin to show up in junk yards in worn-out and wrecked vehicles. For high-performance purposes, these new-style engines should be avoided because the design changes make them incompatible with high-performance parts.

Only the newer-style valve covers will fit the heads as well as a newer-style manifold. Since we will not use stock heads or manifolds these items are not an important point, but pre-1986 style flywheels will not fit the crankshaft on the newer engines. Performance flywheels and clutches are designed, so far, for the pre-1986-engines. As time progresses, no doubt there will be performance parts for these engines. But, at the present, parts are limited or nonexistent. Also, a newer-style engine precludes putting on used parts that may be bought at a bargain. These are all things to consider when on a budget.

The rear of a 1986 and newer block. Note the dowel, stud, and bolt holes on the machined part. An aluminum plate goes here and holds a one-piece oil seal.

Four-Bolt vs. Two-Bolt Block

How about a four-bolt main block and a steel crankshaft? That combination seems to be synonymous with increased horsepower. Actually, neither makes more power, but is able to withstand the rigors and tortures that lots of horsepower places upon an engine. For the type of engines we are building here, a two-bolt main block and a cast crankshaft are sufficient.

Realistic figures for the type of engine you can build with the parts described in this book are about 300hp for the 350ci engine and 375hp for the 400ci engine on gasoline. When you begin to push the power figure beyond this point, which *can* be done with an expensive set of cylinder heads, the stresses placed on the lower end of the engine demand stronger parts—unless you enjoy picking up pieces of crankshaft off the track.

If you happen to acquire a four-bolt main block and a steel crank at no extra cost, it'll give you a little added insurance. Just don't pay extra money for these items since they are not needed at this stage of the game.

Teardown

You've got your "new" engine home and are itching to begin work. Well, get out your tools and prepare to get dirty. That old grease-caked piece of iron has to come apart so that you can inspect the parts and prepare them for machine work.

An engine can be torn apart on the floor, as we are doing; an engine stand is not necessary. Just make sure all of the oil is drained out of the pan. If it isn't, you will have a giant oil slick on your garage floor. Also, with the water pump removed, tip the engine forward and drain the coolant out of the block. This can be another source for making a big mess on the floor.

After draining all liquids, remove any accessories, such as water pump, power steering pump, valve covers, carburetor, and anything else bolted to the engine.

Once these items are out of the way, begin tearing down the engine itself by first removing the intake manifold. There are twelve 3 8/16in bolts with 9/16in heads holding on the manifold. Remove all of these and lift off the manifold.

The cylinder heads come off next. Each head is held on by seventeen bolts. Remove them all! One way to make sure you have them all is to count them. If there aren't seventeen bolts lying on the floor or on the bench, check again to make sure they are all removed. If not, you will be prying and straining to remove the heads. And yes, if a bolt is left in place, it is possible to pry so hard the head will crack in two or the block will break. In fact, it is not only *possible* but it has even been done on occasion. Be more observant than this!

One side note here: Although it is easier to use an impact wrench to remove many of the bolts on the engine, it is not recommended that you do so. The reason is because the force of the impact may weaken the bolts, thus leading to problems when they are reinstalled. This is not a common problem, but the potential for trouble does exist.

The lifter valley of a 1986 and newer small-block. Notice the three bolt bosses in the center of the valley. Also, the tops of the lifter bores are machined.

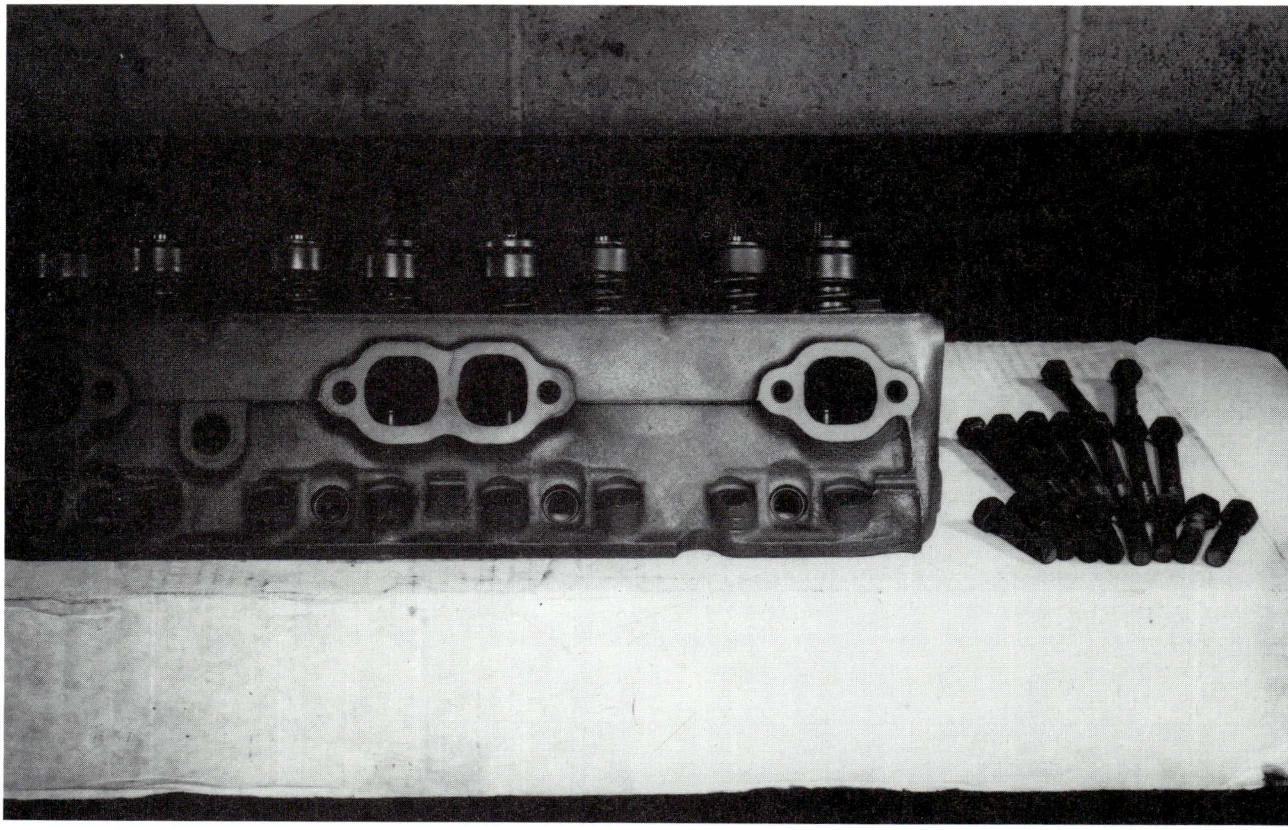

When removing the cylinder heads, make certain that you take out seventeen bolts per head. If you do not count seventeen for each head, double-check to see which one(s) you missed. Some may be hiding underneath all the caked-on grease and oil. If you miss a bolt, it will be difficult, but not impossible, to remove the head; only thing is, it comes off in pieces. You can even break the block by prying a head that still has a bolt holding it on. It has been done!

After removing the heads and the pushrods, turn the block over. Now is when you will discover if the oil and coolant have been drained. If not, have some oil dry or a bunch of old rags nearby to soak up the mess.

The harmonic balancer should be removed next. The *only* way to do this is with a special harmonic balancer puller. If you don't have one, buy or rent one. You *cannot* beat the balancer off with a hammer without damaging it. If a pulley is still bolted to the balancer, remove it. Then select the appropriate bolts from the puller kit to fit the threads in the balancer. The threads are 3/8in fine thread. Be sure to engage the threads in the balancer full depth or else you may end up stripping them, leaving you with the damper permanently attached to the front of the crank. Also, to avoid damaging the threads in the end of the crank, use the protective cone that comes with the puller, or reinstall the balancer bolt. Now you are all set to pull the balancer safely.

After the balancer is off, remove all of the pan bolts and take off the pan, exposing the crankshaft and rods. Remove the oil pump.

Before you proceed any further, you will need a means of marking the main caps and the connecting rods so that they may be reinstalled in their proper order. This is a must! Main caps and rod caps *cannot* be mixed up.

The best method is to use number stamps, although the appropriate number of dots punched with a center punch will suffice. The main caps should be numbered one through five, beginning at the front of the block. Each connecting rod *and* its corresponding cap must be marked, too. Again, begin with the rod assembly nearest the front of the block and work to the rear. Failure to mark mains and rods will lead to confusion when assembly time comes. If the caps are installed incorrectly, bearing failure will result. So, again, whether you use number stamps or a center punch, mark the main caps and the connecting rods and caps before proceeding any further.

Piston and rod assemblies are the next items to come off. Once the rods and caps have been

Number the main caps with a number stamp before removing them. They absolutely must go on in a certain order. Chevrolet does not always number their caps at the factory, so it is up to you, the rebuilder.

numbered, remove the rod caps one at a time. The caps can be stubborn sometimes. The best way to break them loose is to strike them with a hammer and a brass punch to avoid damaging the bolts. Do *not* pry the caps with a screwdriver! This will damage the caps and lead to a short engine life.

Once the caps are loose, protect the crankshaft from rod-bolt damage by covering the bolts with special protectors or with a short piece of hose that fits snugly over the bolts. Now drive the piston out of the block, from the bottom side, by tapping it through with the wooden handle of a hammer. Make sure the rod doesn't hang up on the bottom of the cylinder. With one piston and rod assembly out, you will have just seven more to go. Follow the same procedure, rotating the crank to position the rod for easiest exit.

As far as the ridge reamer is concerned, it is a tool of the past! Do *not* use one. You may easily do damage to the cylinder with a ridge reamer. If the pistons do not come out of the bore, since you will replace them anyway, damage them if necessary to remove them.

Remove the main caps, after numbering them appropriately, and lift out the crankshaft. The block is now bare and ready for cleaning and inspection.

What to Look for When Disassembling

When taking the engine apart, don't fling pieces left and right. Take your time and carefully inspect the parts as you go. If there is something wrong with the engine, it will show in both wear patterns and how the engine comes apart. Be alert to the signals.

One thing to note as you go

The rods and their caps have to be numbered, too. Each cap must stay with its respective rod. Therefore, number both the cap and connecting rod on the same side so that the numbers are next to each other.

Remove the rod bolt nuts and with a brass punch and a hammer, knock the rod loose from the cap. If the cap comes off hard, or if it slides freely and then becomes tight again, it is because the bolts are tipped. This indicates that the big end of the rod has stretched. If the cap comes off smoothly, chances are good that the rod big end is within specification and will not have to be rebuilt.

along is how the rod caps come off. If the caps don't slip off but become increasingly difficult to remove as you slide them over the bolts, then the rod big end is probably stretched. When the rod end stretches, it goes from being circular to elliptical. Stretch a circle and it has to get smaller across its middle. This is what the end of the rod does, which in turn causes the rod bolts to taper in toward each other. Thus, the caps cannot slide off easily.

What all of this means is that when the rod is rebuilt, too much material will have to be cut from the cap and the rod to allow re-honing to size. The result will be a short rod that won't allow the piston to come up high enough in the cylinder, resulting in less compression. The rod rebuilding procedure will be explained in more detail in chapter 3.

Something else to note about the connecting rods is that each end of the rod has a pad; the most prominent one, and the one you can see with the piston still attached, is the one on the big end. This pad is for factory balancing.

The rods are balanced according to weight—each one should weigh about 600 grams. The rods come out of the mold and are machined and weighed. Then the pad is cut until the proper weight is obtained. If a lot of pad has to be cut, that means the bulk of the weight is in the rod beam. If only a little of the pad needs cutting, then the bulk of the weight is in the pad. You want a rod with most of the material in the beam where it will add strength. So, look for rods with most of the balance pad cut away.

After removing the rod cap, place the protective boots over the bolts and drive the piston out of the top of the bore. Use a soft-handled hammer to prevent damage. The boots over the bolts are not for protecting the bolts, but for protecting the crank from nicks in case the bolts hit it. No matter how careful you think you can be, use the protectors; a deep nick in the crank can be costly.

The main bearing caps should fit snugly. There should be a slight press fit between the cap and the block. You will probably have to give the caps a tap, as shown, to free them. If a cap is loose in the block, that is, you can move it from side to side, better look for another block. The cap has probably been stretched and beyond repair.

You can also get a pretty fair idea of the consistency of rod lengths by measuring them with a dial caliper. Rod lengths are actually measured from the center of the big-end bore to the center of the small-end bore, but since this is impossible without a special fixture, the dial calipers will suffice. The idea is to get a complete set of rods that measure equal, or nearly equal, in length. A variation of only 0.001–0.002in is desired. If possible, that is if the rods from your engine check out accurately, keep them mated with their crankshaft journal. This is not critical, but may save time down the road. The reason is side clearance, or the amount of room between a pair of rods when they are bolted to a crank journal. If you keep the rods and crank mated, you are reasonably assured of correct side clearance. (More on this in chapter 6.)

As you work, notice how each bearing cap comes out of the block. There should be 0.0045–0.0050in press fit between the cap and the block. The cap should come out with some resistance. If it is loose and lifts right out, this indicates vertical crankshaft movement. This vertical motion can be caused by severe detonation, which places an excessive load on the crankshaft. Worn bearings can be another cause of vertical crankshaft motion. The excessive clearance simply allows the crank room to move up and down. This vertical action beats against the main caps and actually bends them.

If you discover loose caps in your block, you have two options: align bore or hone the block, or find another engine. (More on align boring and honing in chapter 3 as well.)

A bearing should fit tightly in its cap; if it falls out when you turn the cap over, it indicates that the oil film between the crank and the bearing has been lost at one time. When this film is broken, the crankshaft rolls the bearing just like running a rolling pin over a thin strip of metal. It has a curling effect. When the bearing is flattened, it looses its press fit in the cap. Also look for uneven wear or excessive wear on bearings.

That greasy old engine is just about bare now. All that is left is to lift out the crankshaft. Be sure to have some kind of container for the bolts and small parts as you remove them. It is easy to misplace them.

Once the crankshaft is out of the engine, examine it closely. Look at the threads in both the front and rear. If they are stripped or damaged, they will need repair by heli-coil. Also look closely at the keyways on the front of the crank. If they are wallowed out, thus allowing the keys to shift around, the crank, for practical purposes, is junk. Look for another one.

Push the lifters up just enough so that the cam lobes will clear and pull out the cam. To remove the lifters, push them through the bottom of the bore. This prevents the bottom of the lifters from damaging the bores, which they may do if burred or otherwise damaged.

Examine the block itself. Look into all the bolt holes to make sure that they are not stripped or do not have broken bolts. The starter bolts can be a special problem on small-block Chevys.

Besides looking for stripped threads and broken bolts, look for cracks as well. Engine kickback when starting can jolt the starter enough to crack the block. If the wrong bolt is used in the starter, for example, one with a shoulder that is too long, the block can crack when the shoulder is forced against it. You change the starter, lose a bolt, and, without thinking about shoulder length, you use another bolt that fits the threads. Next thing you know, the block is cracked when you tighten the bolt.

Another problem to look out for is broken starter bolts that someone has tried to drill out. Lying on the ground in the backyard, drilling upward, is not ideal work conditions and, in fact, may lead to crooked drilling or drilling into the block itself. A block that is cracked or has been drilled into is beyond repair—you may as well pitch the engine.

Examine the head bolt holes and run a tap down through them. The thread size is 7/16in—14 UNC. If the tap cuts metal, *stop and get a different tap!* Some taps, though they are the correct size, are a bit big and will actually remove metal, making the threads too large. It usually pays to spend a little extra money and buy a high-quality tap. All you will need to do is remove any rust or dirt from the threads.

Also check to make sure that the bolt holes have a chamfer at

Feeling for wear in the cylinders. This method can give you a general idea of the condition of the engine. If there is only a very little ridge you may be able simply to rering. The front cylinders are usually the ones with the most wear since they run cooler; friction increases as temperature decreases.

the top. If they don't, put one there with a chamfering tool or have a machine shop do it. The chamfer prevents the threads from pulling up, which results in improper torque and an uneven deck surface.

After all of the parts are out of the engine, remove the frost and oil galley plugs. This is necessary for thorough block cleaning when it goes through the hot tank at the machine shop. It will require a special tool to remove the rear oil galley plugs, but if you can get them out yourself, you can check for core shift.

When you sight down through the oil galleys you can actually see the shift. It is visible at the point about midway down the galley where there is a restriction, where the galley has a mismatch or shift. The greater the mismatch in the galley, the greater the core shift.

So, how does core shift affect you? It means that the cylinder walls are thinner on one side than on the other instead of being of uniform thickness. If a cylinder wall is too thin on one side, especially after being bored 0.030in, it may be so thin that it will cause cylinder wall flex and hot spots. The walls may even crack, ruining the engine. However, this is generally not a problem at the power levels we are talking about.

The next step is to knock out the freeze plugs. They have to come out anyway, and if you do this yourself you can check the water jacket for damage. A certain amount of rust inside the water jacket of the block is inevitable, but if it is coated with heavy scale, then toss the block. Thick, scaly rust deposits will not let the engine cool properly and will lead to other problems that you can do without. The thick buildup will hinder the transfer of heat from the cylinders to the cooling system, and there is also a danger of the rust working loose and plugging the radiator.

Cracks in the block will render the part useless. Small-block Chevrolet blocks most often crack in the lifter valley or around the freeze plugs. Sometimes you can detect a crack yourself, thus saving the expense of cleaning and Magnafluxing at the machine shop. A crack on the outside of the block around a freeze plug or drain plug or just below the deck will show as a line when highlighted with dried antifreeze. Anything that looks suspicious along the outside, especially in water jacket areas, should be examined closely.

In the lifter valley, cracks can be detected by checking for brownish areas or for milky white spots if water has been leaking. Anything that looks out of the ordinary needs to be scrutinized. Not all cracks are visible to the naked eye, especially if they are not yet deep enough to cause water leaks, but if you are observant and inspect your block in detail, you may spot them. The sooner a crack can be found, the more money and time you will save. Thus, our motto is, spot a crack, stop immediately!

While inspecting the lifter valley for cracks, also note each lifter bore. Make sure none of the bores are wallowed out—this is rare, but certainly possible on a small-block Chevy. A lifter free of varnish and sludge makes checking easier. If the lifter fits the bore without excess play, then the bore is all right. If it is sloppy in the bore, for all practical purposes, the block is junk. If the lifter won't start into the bore, look for a burr and remove it with a wheel cylinder hone.

One detail that is easy to overlook and may not even seem important when purchasing an engine is, which side of the block the dipstick tube is on. By all means, get a block with the stick on the left for two reasons: oil pans are much more common for the left-hand-stick block, and if the engine is going in a circle track car, the oil will not fling out the dipstick tube in the corners.

Ready for the Machine Shop

The block has been stripped down and all the pieces are laying on the shop floor. The pistons and rods will still be assembled, as will the cylinder heads. These are items that you may want to leave for the machine shop, unless you have a press for the piston and rod assemblies and a valve spring compressor for the heads. Most machine shops do not charge for disassembly; it is all part of the cost of a rod rebuild and a valve job.

A good shop will inspect the parts, too, just as you did when you took the engine apart. There is no extra charge for this, so be more than willing to let them look over your pieces—their trained eyes may find things that you have missed.

A good machine shop will have equipment you don't have, such as Magnaflux for checking for cracked parts, rod straightness checkers, micrometers, and dial bore gauges. Only with this expensive, precision equipment can a definite evaluation of your engine parts be made. Your job is to look for major problems as best you can, as you disassemble. Usually, if a problem is easily detected, the damage is too severe for practical reconditioning.

But before you take your parts to the machine shop, you will need to make some decisions. You will need a detailed plan, a road map if you will, to follow to reach the correct end result as effortlessly and painlessly as possible.

This plan involves answering the following questions: What is the intended application for your engine? How much compression do you want? What kind of cylinder heads will you use? What kind of pistons will be needed? The combinations are many; in the next several chapters we will present the most practical ones, combinations that will be least expensive, yet reliable and produce respectable amounts of horsepower.

Selecting Components

3

Before you take your disassembled engine to the machine shop, you must make some decisions. The major decision is, What will be the purpose of the engine? Will you use it exclusively on the street; will it see duty as both a street and drag strip engine; or will it be used only on the track? The intended use for your engine will affect the parts you put into it and subsequently, some of the machine work to be done.

Let's look at each of these use categories, one at a time.

Engines For Street Use Only

If your car will see street use only, there are two critical issues to consider: one is compression, the other is cam selection. First, compression.

With the octane levels of today's gasoline, 9.0:1 compression is the absolute upper limit. Even at this point you will be right on the edge, so a little less will be better for the good of the engine. The danger with too much compression is the possibility of detonation.

Detonation occurs when the fuel explodes rather than burning at a controlled rate. To understand what happens under detonation, imagine a little person inside the cylinder hitting the piston with a sledge hammer—the rings may be shattered and the piston may be destroyed, rendering the engine useless. Repairs will be expensive. So, keep the compression at a reasonable level; high compression is for the racetrack only.

The cam for the 355. This hydraulic grind from Elgin is similar to the old 350hp 327 cam from the sixties. It is very streetable and is compatible with all vacuum-operated accessories, a stock torque converter, and has a good idle with only the faintest lope to it. When you mash the throttle, though, this thing comes to life. It produces excellent low- and mid-range torque—just where you need it for the street.

Selecting a Camshaft

A camshaft may make or break an engine for street performance. A "big" cam may sound impressive at idle, but when you step down on the accelerator and the engine bucks and bogs, that's not impressive. A big cam cannot be used for smooth starts at low rpm—where the majority of street driving takes place. In order to get the engine rpm up where the cam will work a high-stall torque converter, a clutch, or lower gears must be used.

Consider, too, if the car has a clutch, you will have to bring up the rpm before releasing the clutch, which no doubt will result in tire-squealing starts. Not the thing for the street!

Also, a big cam, that is, one with lots of valve overlap, will create low manifold vacuum, meaning if your vehicle has power brakes or any other vacuum-operated equipment, it will not work. *Think* about all of these factors when selecting a cam for the street.

For street use, a hydraulic camshaft is the best choice. This means the lifters that are used with the cam are self-adjusting hydraulic types. Once they are set when the engine is assembled, they should remain set for the life of the engine. This may be an important consideration if you have a car with such close quarters under the hood that removing a valve cover is a major ordeal (the valve covers have to come off to adjust the lifters).

Another factor in favor of the hydraulic camshaft is its ease on the valvetrain. Without any lash to take up before the rocker arm opens the valve, parts don't get as severe a workout and should

These Silv-O-Lite pistons are stock replacement parts, complete with the dish to lower compression. By using these pistons in the 355 we kept cost down and duplicated a stock lower end, just as it comes from the factory.

Notice the difference in design between the new-style, left, and old-style, right, pistons. The upper edge on the new piston is chamfered, leaving less distance between the top ring and the piston top. There is less "dead space" above the ring where the air-fuel mixture can be hidden between the piston and cylinder wall where it won't burn. The primary reason for this is to reduce emissions. The secondary benefit is that the engine is more efficient, since it burns the mixture more completely; this translates into more power.

Parts Costs for a 355ci Engine

Cam bearings	$16.50
Core plug kit	$3.00
Sealed Power kit (rings, bearings, Fel-Pro gaskets)	$100.10
Silv-O-Lite 1470 pistons	$80.00
Elgin timing gears	$20.50
Elgin E903-P cam and lifters	$97.00
Elgin stock oil pump	$20.50
Elgin LT-1 springs	$25.00
E-Loy 1.94in intake valves ($4.50 each)	$36.00
E-Loy 1.50in exhaust valves ($4.50 each)	$36.00
World Products Dart 305 heads	$570.00
Weiand 7502 manifold	$70.00
Total	**$1,074.60**

live longer. Also, the hydraulic cam is quieter than a solid type because of the lack of lash.

When choosing a camshaft, refer to the cam manufacturer's manuals and call the tech line number if you have any questions. The cam we have chosen here for the 355ci street engine is an Elgin, part number E-903-P. The Elgin is a hydraulic stick and will work well for this engine. The cam has a power range of 2000–4800rpm, producing good mid-range torque from about 2400–3200rpm. It has a fair idle with a slight lope, meaning that your buddies will know you have a real "cam" in the car, yet it will quickly smooth out and work at low rpm. It works well in a stock or mildly modified engine with compression below 10.3:1, and produces enough manifold vacuum to run all vacuum-operated accessories. The Elgin E-903-P can be used with a stick shift or an automatic with a stock torque converter. The cam is even good for mild bracket racing.

Selecting Pistons and Rods

The pistons for the 355 are stock cast replacement types. They are inexpensive and will keep the compression down to tolerable levels. By using these pistons, we can demonstrate what you can do if your engine does not need to be bored, or if you simply want to bolt on some "go fast" pieces. You can add the

The dual-plenum Weiand 7502 or a similar manifold is an economical and effective manifold for the 355ci engine. Its power range closely matches that of the 903-P cam. A dual-plenum design is a must for good response on the street. It works from idle on up to about 4800rpm.

cam, manifold, and heads as described here and achieve similar results.

The pistons we are using are Silv-O-Lite stock replacements, part number 1470. Shop around and look for the best deal; these are simply cast original equipment manufacturer (OEM) replacement parts.

The connecting rods are completely stock. No polishing the beams or high-strength rod bolts or any of that "trick" stuff. Stock rods are fine for up to about 400–450hp as long as rpm is under 7000. So be realistic.

Selecting an Intake Manifold

The manifold for this combination is a Weiand dual-plane design, part number 7502. The one we will use is a non-exhaust gas recirculation (EGR) type, but if the laws in your state require your vehicle to be EGR equipped, then buy an EGR manifold. All the manifold manufacturers offer both styles.

This particular manifold is good from idle up to about 5500rpm, and matches nicely with the range of the camshaft to create a well-balanced package. All components in the engine must be compatible or you will end up scratching your head, trying to figure out why the thing isn't making any power. It should be obvious that if you had a high-rpm manifold on the engine with the low-compression and the E-903-P camshaft, the parts in effect would be working against each other. The manifold would just begin to come into its efficiency range at about the same time the cam would fall off. The parts have to be synchronized to produce maximum horsepower.

Selecting an Oil Pan and Pump

Stick with a stock oil pump for street use. Forget all about any notions of high-pressure pumps or high-volume pumps—they simply have no place here. Figuring about 10psi of oil pressure per thousand rpm, which is a general rule of thumb, only about 50–55psi are needed for street use. A stock pump produces about 40–45psi—plenty for an occasional burst of speed.

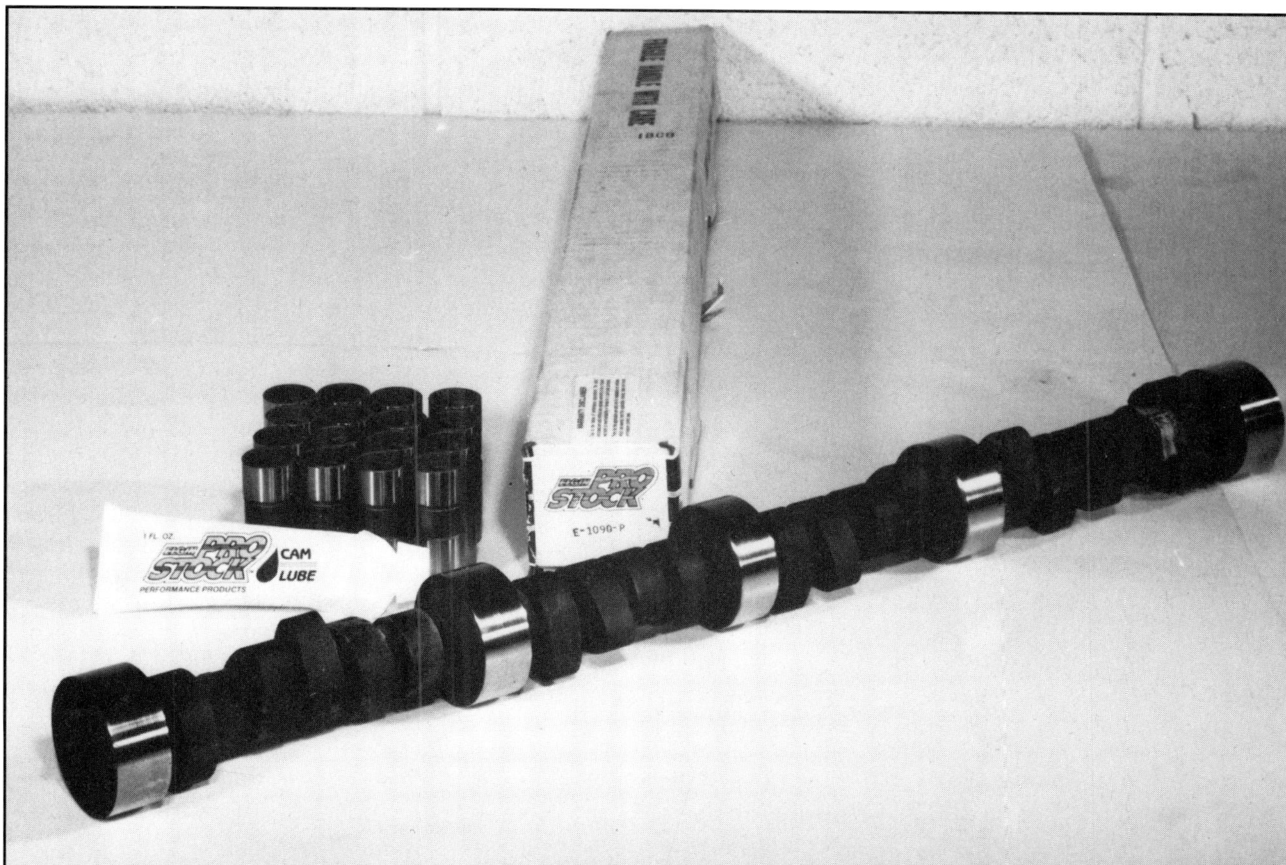

The 1090-P solid-lifter camshaft from Elgin is a serious race piece. It has a rough idle and an effective power range of 3200–6500rpm, and is not compatible with vacuum accessories, which a race car does not have anyway. The solid lifters require the valve lash to be adjusted periodically, meaning you must be willing to tinker with the engine.

The pistons for the 406 are Keith Black Signature Series from Silv-O-Lite. At $260 for the set they are an economical, yet durable racing piston. With the 67cc Dart heads, these domed pistons yield 12.0:1 compression.

Parts Costs for a 406ci Engine

Cam bearings	$16.50
Core plug kit	$3.00
Sealed Power kit (rings, bearings, Fel-Pro gaskets)	$114.40
Silv-O-Lite Keith Black Signature Series Pistons KB-127	$260.00
Elgin timing gears	$20.50
Elgin E1090-P cam and lifters	$135.00
Elgin high-volume oil pump	$33.50
Competition Products #327 valve springs	$66.55
World Products Dart S/R heads	$425.00
E-Loy 2.02in intake valves (6.00 each)	$48.00
E-Loy 1.6in exhaust valves (7.00 each)	$56.00
Elgin hardened pushrods	$20.00
Competition Cams 1.6 ratio roller-tip rockers	$120.00
Poly-Locks	$18.00
Champ pan and pickup	$133.11
Guide plates	$15.00
Weiand 7547 manifold	$70.00
Total	**$1,554.56**

Using a high-volume pump with a stock pan can be dangerous to your engine. It is entirely possible to pump the pan dry at high rpm. When the pan empties, the pump will have only air to move and since air does not lubricate well, the engine will be destroyed. So, stick with what the Chevrolet engineers designed when it comes to the oil pump.

As with the oil pump, a stock oil pan is fine for the street. All that extra oil in a deep pan is not needed; with a special pan, a different pump and/or pickup has to be used, and all of these things cost money.

If there is any room for doubt here, stock oil pans are used in the more economical forms of racing: hundreds of street stocks and modifieds compete with stock oil pans and these cars are accelerating quickly, braking hard, and slamming around corners. And the engines live. What you do on the street can't even come close to that kind of abuse.

Engines For Racetrack Only

If the engine is going to see only racetrack use, then there are more doors open for modification. The car will have a high-stall torque converter or a stick shift; idle quality is totally unimportant; compression can be much higher if racing fuel or alcohol is used; and there are no vacuum accessories to worry about. The objective is all-out horsepower.

As mentioned, this is an economical engine, and it cannot compete with a pro engine with aluminum rods, special crankshaft, and a $10,000 induction system. But, for bracket racing, street stocks, or modifieds where a claim rule is in effect, this engine should hold its own very well—plus it is an economical way to build power.

Just as compression and camshaft are critical in a street engine, they are key elements in a racetrack engine as well. Compression makes horsepower, so the more the better—at least up to a point.

Selecting a Camshaft

The camshaft for a race-only engine does not have to exhibit good idle characteristics since the car only idles when in the pits. On the track, whether it be drag strip or circle track, the rpm will already be up in the power range when the engine is called upon to do its thing.

A solid-lifter camshaft is a fine choice for a race vehicle since it will work at higher rpm; the valve covers come off easily for lash adjustment; the clatter can't be heard over the open exhaust; and since the car should be maintenanced after every race, setting the lash is simply part of the routine.

The camshaft chosen for the 406ci is an Elgin solid lifter, part number E-1090-P. It exhibits rough-idle characteristics with a heavy lope. Power range is from 3200–6500rpm, with a mid to high torque range of about 3800–6500rpm. The cam works well with a compression ratio of 10.5:1 to 12.0:1. The engine must be modified—this cam won't work in a stock engine. It should be ideal for either bracket racing or circle track racing.

Selecting Pistons and Rods

The pistons for this engine are a little better than the cast replacements for the street 355; they are Keith Black Signature Series pistons from Silv-O-Lite. Part number is KB-127. These domed pistons use stock width rings and are called hypereutectic performance pistons.

Some explanation needs to be given here. First, how do you pronounce that long word? It is hyper-eu-tec-tic. Now, what does it mean?

Pistons are made from aluminum silicon alloys. These alloys fall into three categories: eutectic, hypoeutectic, and hypereu-

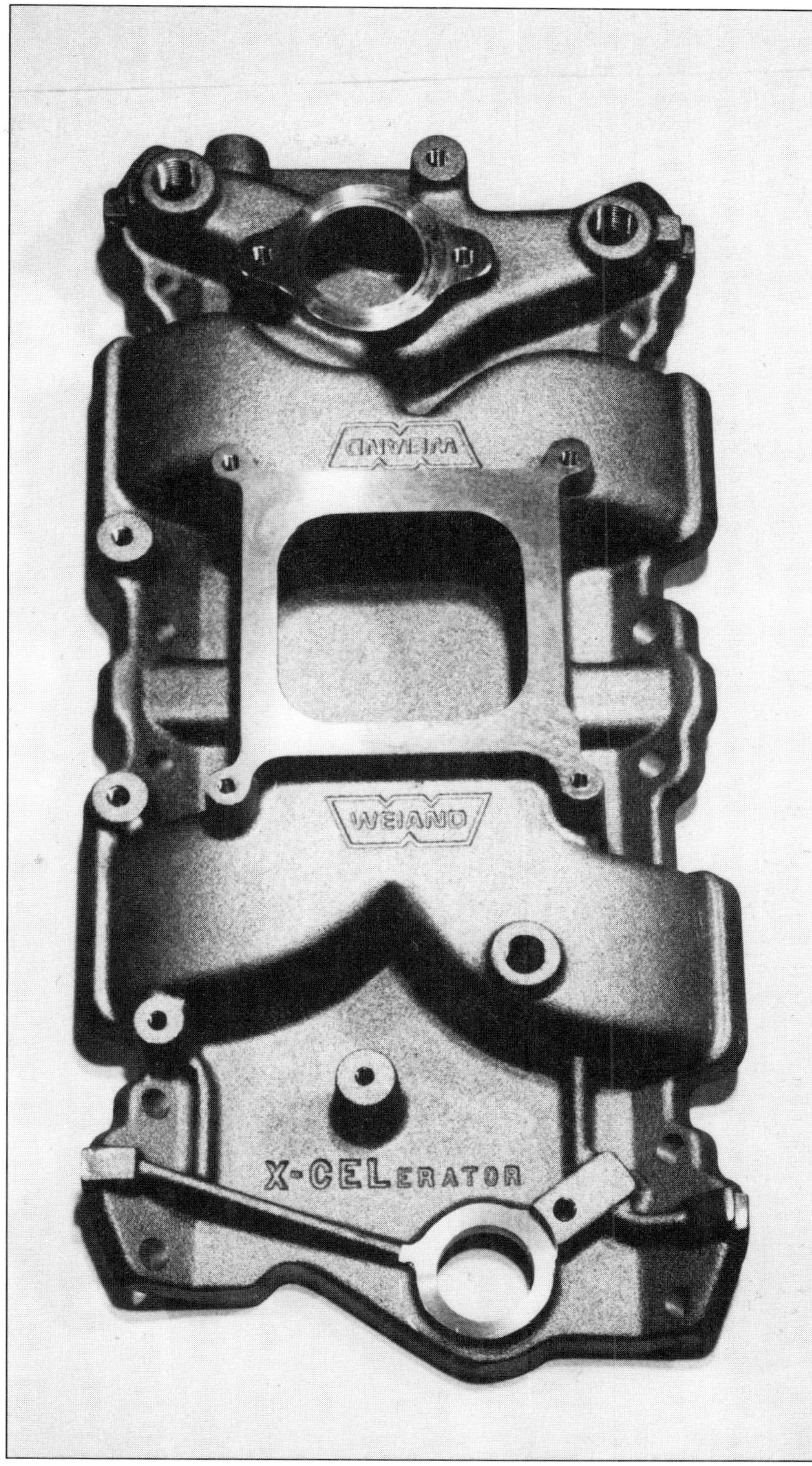

The Weiand X-Celerator, part number 7546. At $70, it is a good manifold for the money for a racing application. The open-plenum design allows it to be effective at higher rpm, though low-speed characteristics suffer. The low profile of this manifold serves two purposes: it is proven effective for making power (taller is not always better), and it will fit beneath a stock hood. The second benefit is good if your class requires flat hoods; it also will save you the time and money modifying the hood.

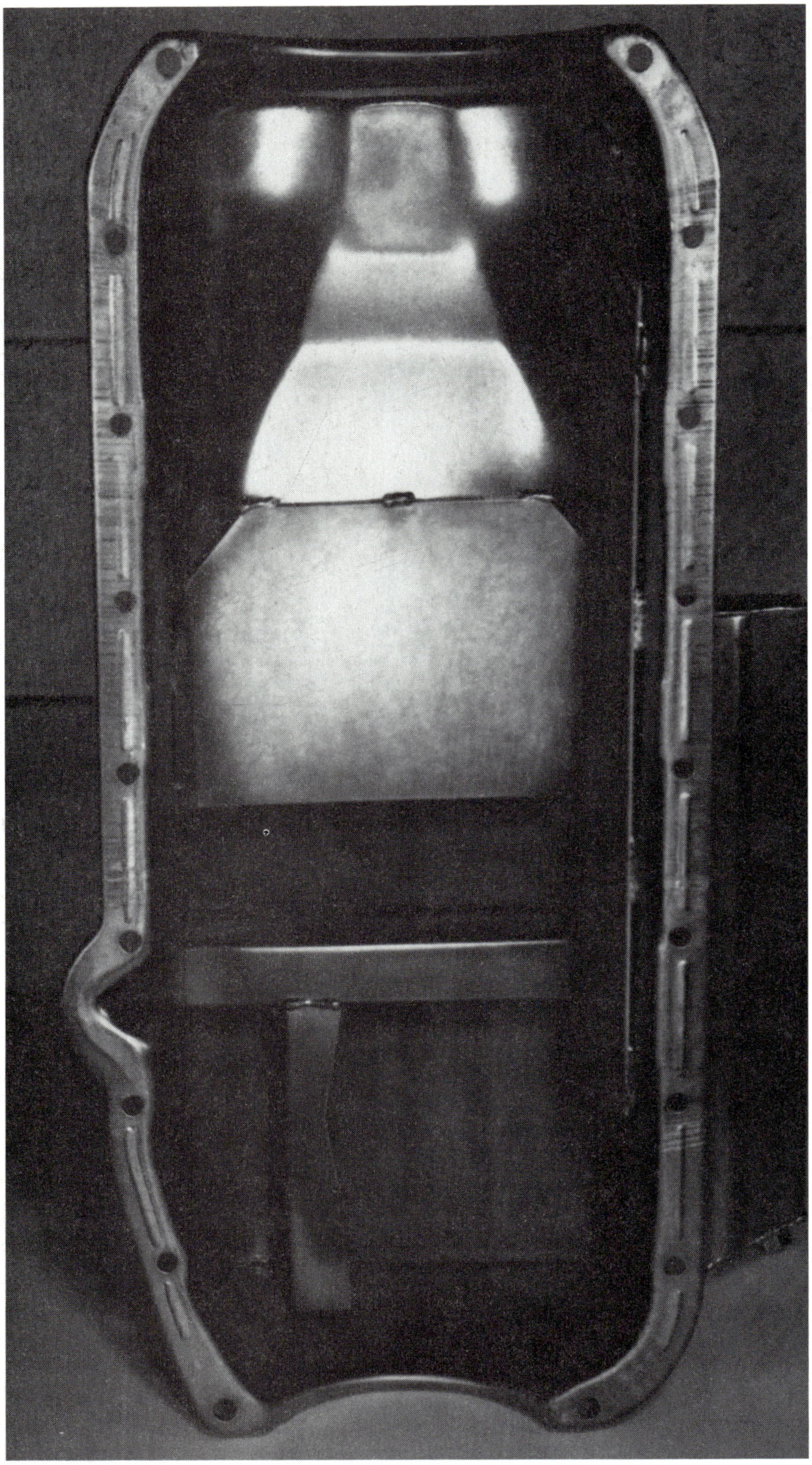

This Champ pan from JR Manufacturing is an excellent circle-track pan. It has an extra capacity sump and baffles to control the oil in cornering, braking, and acceleration. The pan is made of good, heavy-gauge steel, the rails are flat, and the holes line up perfectly with those in the block. It comes with a pickup that is fool-proof to install—it is pre-set for the depth of the pan. At $133, this is a bargain!

tectic. An analogy to clarify these terms is that of a glass of iced tea. Imagine adding sugar to the tea. When the sugar is stirred it dissolves and becomes inseparable from the tea. But, if you keep adding sugar, a point of saturation will be reached. At this point, the excess sugar will fall to the bottom of the glass.

Similarly, silicon is added to aluminum and is dissolved until it becomes inseparable. But, just as with the sugar in the tea, if enough silicon is added, a saturation point will be reached and the excess silicon will drop out of solution in the form of hard, primary silicon particles.

The point of saturation in aluminum is called the eutectic and occurs when the silicon level reaches 12 percent. Aluminum with a silicon level below 12 percent is hypoeutectic; aluminum with a silicon level above 12 percent is called hypereutectic. Most hypoeutectic pistons have about 9 percent silicon. Eutectic pistons have a silicon level of 11–12 percent.

The Keith Black Signature Series pistons have a silicon level of 16–18 percent. This means that they have great scuff and seizure resistance as well as greater strength. Thermal characteristics of these pistons are improved, too, because the silicon particles that precipitate out act as tiny insulators to keep the heat in the combustion chamber and prevent heat transfer to the piston. This results in about 15 percent less expansion than in pistons made of other alloys.

Stock rods are fine for this engine, too. As mentioned previously, if you feel the urge to polish the beams and do a lot of trick stuff to them, throw them out and get good aftermarket pieces. There is no reason to do this, though; the stock rods will hold up quite well even in this big 406ci engine.

Better rod bolts may be a good idea just for some extra insurance, since the power figures will be getting up there. It depends, too, upon how hard you plan to turn the engine; continuous rpm above 7000 may be reason enough for some better bolts.

Selecting an Intake Manifold

The manifold on a race engine, like the camshaft, can work at higher rpm than on a street engine. An open-plenum style works well here. It doesn't have the low-end response of a dual-plane manifold, but works beyond the point where the dual-plane drops off. The one we chose is the Weiand 7547, an economical, but efficient manifold. It should work well with the cam in the 406, producing power in the mid to high range.

Selecting an Oil Pan

You may be tempted to use a stock Chevy truck oil pan on your racing engine because of its greater capacity. Do not do it! The truck pan does not contain the proper baffles for hard cornering and braking. The oil will slosh away from the pickup under these conditions, leaving only air for the pump to move through the oiling system—this is certain death for the engine.

For our 406 we chose a 7qt aftermarket pan made by JR Manufacturing, Inc. This is their Champ Pan, built specifically for circle track racing. It has all of the appropriate baffles to contain the oil under hard cornering and braking conditions.

Selecting a Timing Chain

The timing chain we chose is an economical roller that sells for about $21. This is not to be confused with a *true* roller. There is a lot of difference in quality and cost between the two, with the true roller being the better and the costlier.

A true roller actually has a roller on each link in the chain that rolls over the gear teeth. A roller-style chain lacks the actual rollers. Roller-style chains are common in Chevy truck engines, so they are dependable and long lasting, though they do tend to stretch rather quickly. Your budget will determine what you buy.

Main Studs vs. Stock Bolts

There are stud kits on the market that replace the main bolts. Do not invest in them! They are not appropriate for the applications described in this book. The only reason for replacing bolts with studs would be if the engine will be torn down a lot. If the bolts are removed and reinstalled constantly, the threads in the block will eventually wear out.

Contrary to popular belief, *studs do not add strength.* This is a common misconception, due in part to advertising. Their main purpose is to save threads. When studs are installed in a block there is a danger of shifting the caps, thus distorting bore size. You do not need this kind of headache.

The same advice applies to head studs and the oil pump stud. Stick with stock bolts and save yourself some money and possible hassles.

Selecting Pushrods

Stock pushrods work just fine with valve lift up to 0.500in; we used these in the 355. If the cam has over 0.500in lift, then a stock replacement pushrod with a ball tip has to be used. Without this ball tip, the rocker arm and pushrod will bind.

If your cylinder head has guide plates, then hardened pushrods must be used—the plates will eat up stock, unhardened rods. The hardened pushrods we used in the 406 sell for about $1 apiece. They are not chrome-moly, but are hardened to withstand the friction against the guide plates. The valve spring pressure in the engine is not high enough to warrant the more expensive chrome-moly rods anyway.

We haven't discussed cylinder heads yet; they will be discussed in another chapter since there are so many things to be said about them.

Now, off to the machine shop with all those engine parts.

The timing sets we used are nothing special. Just a good, economical roller style that sells for about $21. Notice the three keyways in the crank gear. These are for changing cam timing; the cam may be installed straight up, four degrees retarded, or four degrees advanced. All you have to do is use the appropriate keyway. This economical timing set should not be confused with a true roller, however; there is no comparison, but a true roller is a lot more money.

Machine Work

4

In keeping with the concept of building an economical engine, the machine work should be restricted to only what is necessary. Forget all the exotic tales you've heard about such things as relocating lifter bores, changing the angle of the valves in the heads, angle milling, and the like. There may be a time and a place for these operations, but not here. Save your money!

Instead, be concerned with getting quality work done on the basic tried-and-proven operations vital to a dependable engine.

Selecting a Machine Shop

The best place to find quality work is your local machine shop. If you do not already have a shop that you take your work to, then ask around and find out which shops have good reputations. Concentrate on reputation, not labor rates. The cliche "You get what you pay for" applies here. A shop that does work for considerably less than its area competitors most likely cuts corners on work. If you pay $50 for a valve job, then you will probably get a $50 valve job, meaning no guide work and single angle seats that are too wide. The engine may run, but not well and not for long.

Visit the shop before you take your engine pieces in for work. Get to know the machinists, and feel free to ask for a tour and an explanation of how their work is done. Ask questions and listen carefully. Remember, the machinist is a professional. You wouldn't try to tell your dentist or doctor how to treat you, so don't tell your machinist how to do his or her job.

You must have clear communication with the machinist; without this, you both may end up frustrated. You will be disappointed in the work done, and the machinist will wonder what kind of screwball you are. So, make sure you understand each other. And don't try to slip in extra work after a quote is given. Be honest about what you want and what you will bring in the door. Don't ask for a price for block cleaning and boring and then drag in valve covers and bolts and expect to slip them in for the original price. If there is any kind of a personality conflict, move on and check out another shop.

Don't try to get the machinist to guarantee that your engine

Get your work done at a local machine shop, one you can go into and visit, instead of a mail-order shop. Ask friends who they have good luck with; stop in and get to know the machinists before you take in your work; and ask questions and explain what you want done, but don't try to tell them how to do their jobs. And don't look at price only. Consider the whole package: reputation, willingness to accommodate you, and personality compatibility between you and the machinists. Remember, a local shop has a reputation to maintain because it depends upon repeat customers to stay in business.

This cylinder head is an example of what you can end up with through mail-order. Notice that all of the springs have the thick rotating retainers on them—the retainers should be only on the exhaust. Also, there are no oil shields under the retainers. These are the obvious errors. When this head was taken apart it was also discovered that one spring was shorter than the others (it was used, the other fifteen were new). It was supposed to come with hardened exhaust seats installed; they weren't. These errors would not have kept the engine from running, but it would not have run as efficiently as it should. The soft exhaust seats, too, would have soon been beaten in. We are not saying that all mail-order work is bad, but why take the chance? You also will face lots of hassles and shipping expense returning the products.

will never blow up; these kinds of guarantees are impossible to make. It all depends upon how you treat the engine, how you assemble it if you do so yourself, and a certain amount of luck. Remember, anything mechanical can break, and you are going to run the engine hard, at least upon occasion. The object is to *decrease* the odds of failure as much as possible.

A good shop *will* guarantee its work within certain specifications and, if the shop does any assembly, will guarantee the pieces to be correctly assembled. Also, reputable manufacturers will guarantee their pieces to be free of defects. These kinds of guarantees should be expected. If there is any doubt, ask the machinist and/or the person who sells you the parts. Just don't expect someone to be responsible for any mistakes *you* make, or any abuse the engine receives while *you* are driving the vehicle.

We stress the importance, here, of using a *local* machine shop; mail-order machine work is risky. The prices may be appealing, but once again, you get what you pay for. Most shops that do volume business cut many corners to keep the work moving along. These shops are run like a factory, with production volume being the number one criteria—get it out the door as quickly as possible. In order to keep the pieces moving quickly, time cannot be spent on careful machine work. Also, not all of the operators are trained machinists, but simply people who do nothing but the one operation. Even though they may be good at their one job, if they don't have an in-depth understanding of engines and know *why* they are doing what they do, they can mistakenly let faulty pieces get by.

Typical examples of problems associated with mail-order machine work include cylinder heads that are not surfaced flat, guides that are too tight, and incorrect retainers; rods that have distorted big-end bores; and crankshafts with no radius on the journals. These things may not hinder an engine from running, but they most certainly will decrease its life and may even lead to broken or damaged parts. Rods with incorrect big ends will lead to early bearing failure; a crankshaft without a radius on the journals may break; a cylinder head with incorrect springs

The oil hole in the cam bearing bores is often too small on the small-block Chevy—just a factory defect that can be cured easily. Drill the hole so it matches the size of the one in the main bearing bores. Have the machine shop do this for you; the charge should be minimal.

and/or retainers will not rev to the rpm that it should. Don't risk shabby machine work through a mail-order shop.

Spend a little more money and get the work done at a shop down the street or across town, one that your friends have had experience with, one where you can walk in the door and see how things are done, one that depends upon reputation for their business. You will be much happier in the long run.

Machining the Engine Block

The machinist will measure the bores of the block with a dial bore gauge to determine the amount of taper and runout from wear. If there is less than 0.005in taper in the cylinders and they are less than 0.003in out of round, you can simply re-ring using the old pistons. If you're on a strict budget, this is a good move. But, if you want the cylinders perfect or if they are worn more than allowed, then the cylinders have to be bored and new pistons installed.

After the machinist determines how much overbore it will take to clean up the cylinders, the pistons should be ordered. Although the pistons are precision made, they should always be measured and the cylinders finished to fit exactly.

Piston size can vary from brand to brand, and even in the same brand. A thousandth of an inch will make a difference, so it is vital that the cylinders be fitted to the pistons. Too tight and the pistons will scuff; too loose and they will rattle and even break.

Cylinder Boring

The cylinders may be bored with a boring bar or with a cylinder hone, such as the Sunnen CK-10. Either method is fine; it simply depends upon which setup your shop has. In either case, the cylinder is rough cut to within a couple of thousandths of finish size. On a CK-10, course stones are used to rapidly cut away material; on a boring bar a carbide tool cuts away the material.

For the final sizing on a CK-10, a finishing stone is used to give the cylinder walls the correct texture for ring life. If a boring bar is used, the final finishing has to be done with a hone. Most shops have a honing cabinet for this operation. The block is placed in the cabinet, and honing oil is continuously flooded over the cylinder wall as the hone is used.

In either case, the finished product is a cylinder wall with a 45-degree cross-hatch pattern of the correct texture to allow the rings to function to capacity. Generally, a 300 grit stone will give the desired finish for rings such as we are using: a moly faced top ring and a cast-iron second ring.

There is a theory that an extremely smooth cylinder wall finish will create less friction against the rings and result in more power. While there may be some merit to this theory, it is not practical for the engines here because the rings will not seal well. When the rings don't seal as they should, it will lead to oil consumption and power loss. So, have the machinist stick to the tried-and-true method.

Boring With a Deck Plate

How about having the cylinders bored with a deck plate? This is a good idea, though not absolutely necessary. If the block is bored without the plate, the bores that are round will not be round when the heads are fastened on. When the head bolts are torqued into place they pull on the block, thus distorting the cylinders.

The deck plate is a piece of steel or cast iron that is bolted onto the block, just as a cylinder head is bolted on. The object is to produce the same type of distortions in the block as do the heads. The cylinders are bored and honed so that they are round for engine assembly. This procedure is most beneficial when the engine is fresh—the bores start out round and the rings will seal better. After a certain amount of running time, the cylinders in a

After the block has been stripped, hot tanked, and checked for cracks and cylinder bore condition, the next step is boring and honing. The Sunnen CK-10 shown here bores the cylinders with stones. It is a fast, accurate means to bore a block. The finish is put on the cylinder walls by a set of finishing stones of the grit recommended by the ring manufacturer.

If the block needs to be line honed, this is one method. The CK-10 has a line-honing unit attached. The main caps are cut in a rod-and-cap cutter, torqued in place, and the block is set up in a fixture. The main bores are undersize after the cutting operation; a mandrel with honing stones on it is run back and forth through the main bores until they are back to size.

block bored without a plate will become round and the rings will seal.

It does cost extra money to have an engine bored with a torque plate. Done properly, it is more of an operation than simply bolting it on. The bolts must project through the plate and into the block to the same degree that the head bolts project through the head and into the block. Attention to this kind of detail will ensure that the cylinders are distorted to the same depth as when the heads are fastened on. It takes time to do this, and time costs you money in the shop.

Labor Costs for a 355ci Engine

Cleaning	$50.00
Install cam bearings and core plugs	$25.50
Bore with torque plate	$95.00
Align hone	$80.00
Deck	$80.00
Surface heads	$20.00
Recondition rods	$48.00
Install pistons	$28.00
Total	**$426.50**

Line Boring and Honing

Both of our engines have been line honed. The main bearing bores on all blocks are out of round to a certain degree due to stress and heat. If the main bores in the block are 0.001in or less out of round, then there should be no problem.

Remember when the engine was torn down and a big deal was made of examining the main bearings? Here is where it pays off. This is one way to determine if the centers of the bores are concentric. If the bearings are evenly worn, then the bores are in line with each other. If the bearings are worn heavily on one edge, then alignment is off.

Main bearing bore alignment can also be checked, to a certain degree, with a straightedge and feeler gauge, although there is a certain amount of error involved.

If you want to spend your money for perfection, or if the bores are more than 0.001in out of round and they are not in line with each other, then the block will have to be line bored or honed.

Line boring and line honing are two different methods that

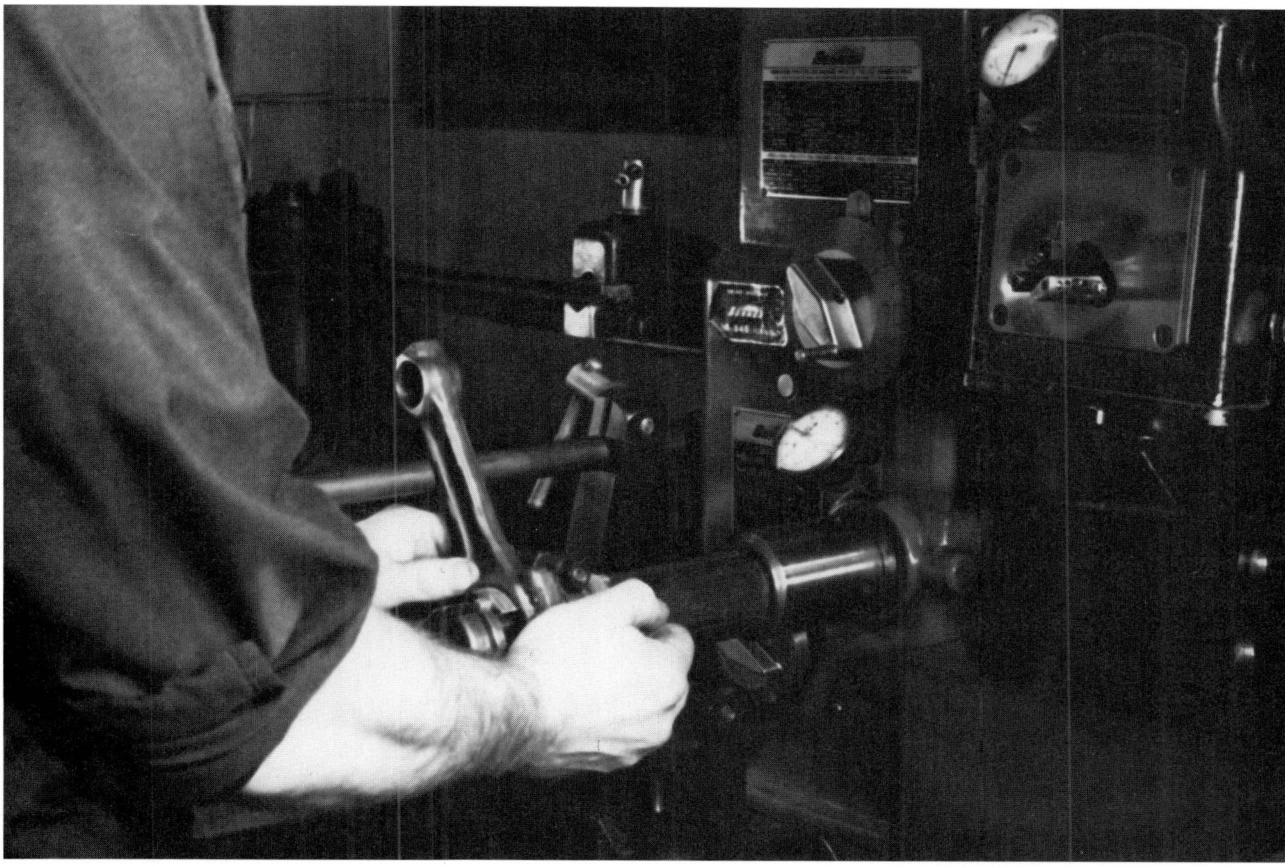

To rebuild the big end of the rods, the bolts are knocked out, the rod and the cap are both cut in a cap cutter, the rod is reassembled, and the bolts are torqued to spec. Just as with the mains in the line-honing operation, the big-end bores are now undersize. The rods are run back and forth in pairs across a mandrel until just enough material is removed to bring the bores back to spec.

achieve the same results: putting the main bearing bores on size and in alignment. Whichever method your shop uses is fine. Our engines were done on a Sunnen line hone, which is a part of the CK-10 machine. This makes it convenient to do the boring and line hone work at the same location.

To line hone the main bearing bores, 0.005in is cut off of each main cap. The caps are cut on a rod-and-cap cutter, which is a special machine that holds the caps, or rods, squarely and firmly in place so that they are cut exactly 90 degrees to their vertical centerline. This is crucial for proper cap fit. The caps are torqued in position in the block and then the bores are honed or bored out to specification. This will raise the centerline of the crank 0.0025in in the block, but is not enough of a movement to create any problems.

Block Decking

There are two instances when a block should be decked. First is if the deck surface is not flat—then it is necessary for gasket sealing. Second is if you want to decrease piston-to-deck clearance to gain compression. If you are faced with either of these conditions, then there is no choice. Otherwise, save your money.

Decks on 350ci blocks generally are in good condition and won't need surfacing. The 400 blocks, on the other hand, often are warped and must be surfaced. It's just a common ailment of the engine.

If you wish to gain compression, block decking is a good way to go. Milling 0.010in off the deck

Labor Costs for a 406ci Engine

Cleaning	$50.00
Install cam bearings and core plugs	$25.50
Valve job (enlarging seats for 2.20in and 1.60in valves)	$75.00
Bore with torque plate	$95.00
Line hone	$80.00
Deck	$80.00
Surface heads	$20.00
Recondition rods	$48.00
Install pistons	$28.00
Grind crank	$65.00
Cut spring seats	$10.00
Machine for guide plates	$48.00
Total	**$624.50**

The only safe way to heat rods. This electric rod heater maintains a consistent temperature, thus preventing overheating which could lead to weakening of the metal. If rods are heated with a torch, there is a real possibility of overheating them.

When the rod is heated, the pin bore enlarges just enough so that the pin can slide through. The fixture on the heater is adjustable so that the pin is pushed in the correct distance. In only a matter of seconds the rod will cool enough that the pin becomes tight, so there are no second chances to get the job done right.

of the block will do much more to raise compression than milling the same amount off the heads. Excessive milling on the heads (or the block) will cause intake manifold mismatch, requiring that the manifold or the intake side of the heads be milled, too. This is just an added expense, and the manifold will not fit another set of heads unless they are milled the same amount.

Rebuilding the Connecting Rods

Have the connecting rods checked for size and roundness on both the small and big end. The small end, that is, the pin end, should be checked first. There should be about 0.0015in press fit between pin and rod. If the bore is too small, it can be honed. If it is too big, it is best to simply throw away the rod and use another one. If the big ends are within specification, use them as they are. If they are out of spec, they have to be rebuilt if the engine is to last at all.

To rebuild the rods, the bolts are removed and the rod and its cap are both cut on the rod-and-cap cutter. Just enough material is removed to clean and square the mating surfaces. It is important that no more than necessary is removed or the rods will become too short. In engine teardown we discussed the importance of examining the rods. If a bearing has spun, the bore will be damaged and too much material will have to be cut from the rod for the honing to clean it up. If you have a rod damaged in this way, throw it away and use another one!

When the machinist cuts the rod and cap and the bottoms are not parallel with the sides, throw away the rod. Again, too much material will have to be cut off, thus making the rod too short. This is one reason to shy away from remanufactured rods; many of them are not cut square and when your machinist tries to correct the problem, he will have to remove too much material.

After the cutting operation, the bolts are reinstalled and torqued to spec. The rods *must* be held in a special rod vise while the bolts are torqued to prevent distorting the rod end. Also, whenever the rod bolts are removed, the rods *must* be rebuilt. The centerlines of the new bolts may be slightly different than the originals, thus shifting the cap. So, don't think that you can simply knock out the old bolt, and put in new ones and be ready to go—it just won't work. You'll end up blowing the engine for sure.

This fixture checks rod straightness. If the rod is not straight, it can be corrected. The rod is clamped in a vise by the big end, and a long, tapered rod is inserted through the pin end which the machinist simply uses as a lever to twist the connecting rod back into position. This operation, though it sounds crude, is perfectly safe and does not weaken the rod.

A side note here on rod bolts: It pays to find an engine that has never been apart before because you know that the bolts have not been overtorqued. If the engine you are rebuilding has been worked on before and you don't know the person who did the work, you may want to install fresh rod bolts. If a rod bolt is overtorqued it will be stretched and weakened. On stock small-block Chevy, bolts go to exactly 45lb-ft of torque; no more, no less.

After the caps are torqued in place, the big end of the rod is now undersize because of the material removed. To bring them out to spec, a rod hone is used. This hone has a mandrel that can be adjusted to specific rod size and uses stones much like a cylinder hone to remove material. The machinist strokes the rods back and forth, in pairs, over the rotating stones until the correct size is reached. The dimensions are checked on a dial indicator on the machine.

Fitting Rods and Pistons

There is only one way to fit rods and pistons: by heating the small end of the rod. By heating the rod, the bore will expand enough that the pin can be slipped through. If you try to press the pin through the rod, galling will occur and the piston may also be damaged.

Don't try heating the rod with a torch, either, to expand the rod. The heat cannot be accurately controlled and if the rod is overheated, it can be weakened due to the structural changes in the metal.

Be sure to have the machine shop put the pistons on the rods to prevent any possible damage. The small amount of money spent will be well worth it.

Grinding the Crankshaft

The machine shop will check the crankshaft for journal size and runout. If the crank is standard size, that is, if it hasn't ever been ground before, and it checks out OK, just have it polished. Polishing is a lot cheaper than grinding.

Standard cranks from Chevrolet usually are on the low side of the journal specifications. Usual-

When a crankshaft is ground there must be a radius put in the corners of the journal. This radius adds strength; a square corner provides a point where the crank can break. More work and time are involved in grinding a crank with a radius; the grinding stone must be dressed with a radius on its edges. This dressing operation takes time and patience. It is a step omitted by some shops, especially the mail-order shops that turn out work on a production-line basis. This radius is a must, though, whether for high-performance or everyday use.

After the grinding operation, the crank should be polished to obtain a smooth finish. This is done with a hand-held belt sander. To get the correct finish the belt must be oiled and dressed—another time-consuming operation. A good shop will do it right.

ly, you can use 0.001in undersize bearings with a standard shaft to achieve the desired clearance of 0.0015–0.002in on the mains and on the rods.

You may figure, Why not use standard bearings and gain a little extra oil clearance? Well, that extra clearance is just not needed and, in fact, is not good. In the old days, when heavyweight oil was used, this extra clearance may have been necessary, but with the big improvement in oils today and the lighter weights that are used, all the additional clearance would do is cause a drop in oil pressure. So don't try any trick clearances; stick to the factory specs.

If the crank needs to be ground, there are some details that a good machinist will attend to. One is journal radius. If the corners of the rod and main journals are cut without a radius, this provides a stress point where the crank can break. A proper radius adds strength, thus lessening the probability of a broken crank. To achieve the radius, the grinding stone must have its edges dressed precisely. This takes time and patience and is a step that is overlooked sometimes, especially in mass-volume shops.

After the crank is ground, the oil holes should be chamfered to remove the sharp edges. This is more for bearing life than oil flow, since the edges can damage a bearing. The chamfering should be done before the final polishing operation.

To get a good finish on the journals, polishing should follow grinding. Done properly, polishing is an extra, time-consuming step, but will add to engine life. The polishing belt has to be oiled and dressed to the right texture to produce the desired smooth finish.

Cylinder Head Prep

5

A correctly machined and assembled short-block, though it will run flawlessly, will not in itself produce power. The short-block will run only as well as the induction system allows it. Restrict the flow and low power figures result; allow the engine to breathe as efficiently as possible and the power figures will increase dramatically.

The heads play a vital role in the breathing ability of the engine. The *right* heads, with some work and a good valve job, can really bring an engine to life. Professional race teams know that the secret of horsepower lies in the cylinder heads. This is just as true at this level of engine building, although the results may be less dramatic.

There are almost an unlimited number of cylinder head options available for the small-block Chevrolet. In keeping with our budget-minded approach, we have chosen the route that should produce the most power for the dollar. We will use Dart aftermarket cylinder heads on both the 355 and the 406 engine. We also will discuss two other options, to demonstrate why the Dart heads present the most economical approach.

Stock Cylinder Heads

Unless your engine is an older high-performance model, it will have large-chamber "smog heads" on it. The large combustion chambers in these heads lower compression to help reduce emissions. Good for the atmosphere, bad for horsepower.

You can use the smog heads if you wish. Domed pistons will

A World Products Dart S/R head, the way it arrives from the factory: machined for screw-in studs, large springs seats for larger diameter valve springs, and machined valve cover rail.

Bottom side of the World Products Dart head. It comes with the valve seats machined, ready for assembly. The exhaust seats are hardened steel inserts for use with unleaded fuel or alcohol. What is not obvious is the thickness of the head. The deck is nearly double that of a stock small-block head. It is also beefier in other critical areas, making it a durable piece that can withstand abuse.

raise the compression to performance levels, but you are still stuck with the ports in these heads, which are not designed for power.

You will be far better off in terms of performance to get rid of them and go to something better. If the heads are not cracked and are in reasonably good shape, you may be able to sell them to a machine shop for their inventory or to someone looking for stock replacement heads for their everyday vehicle.

Double-Hump Heads

Another cylinder head option is to find a set of older castings that originally appeared on high-performance engines such as the 327. These heads can quickly be identified by the two humps cast on the ends, thus the name double-hump heads. These heads may come with either 1.94in or 2.02in intake valves. Of course, the version with the larger intakes will flow a little better. The ports in these old high-performance heads are more conducive to good flow than are the ports in the late-model types. The ports are bigger and straighter, thus offering less hindrance to flow—these old heads indeed were intended to make power.

If you happen to own a set of double-hump heads and they are in good condition, by all means *use* them. But, since these heads are fairly old, chances are they have many miles on them and are worn out. Most likely they will need new guides and valves as well as springs. Also, if they are to be used on a street engine, unless you are willing to use a fuel additive for lubrication, they will need hardened exhaust seats installed. These old heads tend to crack, too, with years of use.

Finding bare castings that are in good shape can be time consuming and expensive. So, being faced with all of these negative factors, a new set of heads with small combustion chambers for compression, good-flowing ports, and hardened exhaust seats, reasonably priced, would be ideal.

Machining Double-Hump Heads

Even though we have used the Dart heads here, we will explain how a set of double-hump heads would have to be reworked in case you have a set handy and prefer to use them. After all, economy is the bottom line and if

Magnafluxing for cracks—the first step in doing a valve job on a used set of heads. Small-block Chevy heads tend to crack across the exhaust seats, especially on the two center seats. A crack will show up as a white line with the Magnaflux. The piece to be checked is magnetized with the electromagnet, and iron powder is sprayed over the area.

Look closely at the exhaust seats and you can see what happens if unleaded fuel is used on soft seats. They are sunk into the head from the action of the valve. Exhaust seats reach extreme temperatures during operation, and the valve rotation will wear them away. The lead in leaded fuel provides a lubricating barrier to prevent this. With unleaded fuel there is no lubrication, so a hardened seat must be used to withstand the friction.

you have a good set of heads to work with, use them!

If the heads are going to be used on the street on pump gasoline or if they are going in a race car that runs on alcohol, then exhaust seat and guide work must be done if the heads are to survive. If they are destined for a race engine that will use racing fuel that contains lead, then the stock exhaust seats and guides may be retained.

The problem with "dry" fuel, that is, fuel without lead, is that there is no lubrication for the valve seats. On the intake seats this is no problem, but, due to the higher temperatures on the exhaust seats, they will be worn right down into the head.

Lifters, pushrods, and valves rotate as the engine runs. It is the friction of the valve as it rotates against the seat that causes the wear; the valve actually grinds away the seat until it sinks into the head. Eventually the valves will be seating so deeply in the head that major problems will result. The stem height (the distance from the spring seat to the tip of the valve stem) will become so great that the geometry will be thrown off and spring pressure will decrease. The seats will become so sunken that flow will be affected. If not corrected in time, the heads may crack or the seats may become unrepairable.

The best method to prevent this problem is to convert the heads to accept unleaded gasoline. The conversion simply involves installing hardened exhaust seats. Unfortunately, it is not a job you can do at home in your workshop; it requires spe-

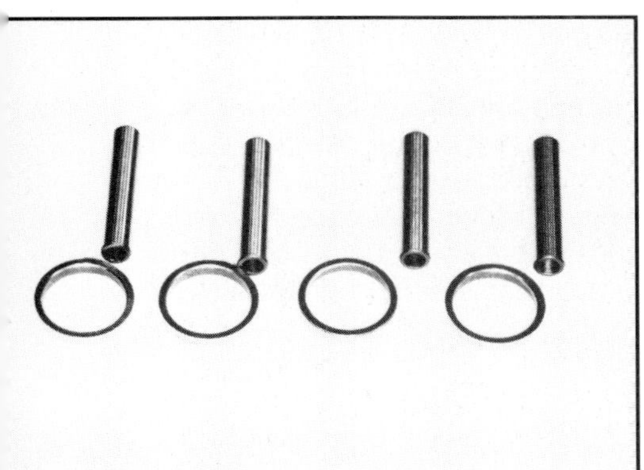

Hardened exhaust valve guides and seats, the parts needed to update a set of old heads, to make them compatible with unleaded fuel.

Guide work has to be done first because the seat work centers off of the guides. The old exhaust guides are drilled and reamed to 0.500in diameter to accept the new inserts.

The new guides are oiled as are the reamed-out old guides to prevent galling as they are driven in. The guides are started into place using a soft hammer to prevent damage.

cial seat and guide equipment. If the heads are off the car and in the machine shop anyway, for a little extra cost this is the time to have the work done.

The heads are first cleaned and Magnafluxed. If they are free of cracks, then the guide work may begin. Intakes, if they are not excessively worn, may be knurled. Exhaust guides, on the other hand, should be replaced with hardened guides. They, too, like the seats, are prone to wear due to the unleaded fuel. The guides we are using in these heads, in addition to being hardened, are grooved for oil retention. The tiny grooves fill with oil and provide a film to help reduce wear between guide and valve stem.

The original guides are drilled and reamed to 0.500in diameter and the replacement guides are driven in. The ends of the replacements are trimmed flush with the originals, resulting in stock appearance with the benefit of a resistant guide.

The bottom end of the exhaust guide, the end that is exposed to the heat of exhaust gases, must be chamfered. The idea of the chamfer is to provide more area at the end of the guide to dissipate heat. If the guide cannot get rid of the heat, it will absorb it, possibly causing a valve to stick. Providing a chamfered area is like making the end of the guide bigger, offering more area for the heat to spread out, thus preventing it from traveling up the guide.

Once the guide work is done, the seats can be installed; all seat work centers off of the guide. The replacement exhaust seats are simply rings made of a hardened material. The original seat in the head is cut out, or counterbored, to a size approximately 0.005in smaller than the seat ring. The ring is then driven into place with a hammer and driver. The press fit holds the new seat securely.

After the rings are installed, the exhaust and intake seats are ground. On the Serdi seat and guide machine, which we used, the seats are actually cut with carbide cutters that do all three angles at once (45-degree seat with a 30-degree top cut and a 60-degree bottom cut). After the seat work is finished, the heads are assembled and surfaced. Now they are equipped to withstand the rigors of unleaded fuel.

If the engine is going to be run on alcohol, all of the original guides should be replaced with bronze guides. Even the hardened guides will not stand up to alcohol. The best bronze guides to use are those that have a 0.500in outside diameter. They are installed just as regular replacement guides are, by drilling and reaming the head and driving in the guide.

Dart Heads

Dart heads are made by World Products, Incorporated. They are a direct bolt-on replacement for stock Chevrolet heads and are made in cast iron and aluminum in a variety of configurations. There is a Dart head for nearly any application.

We chose two of their economical cast-iron versions in the interest of keeping cost to a minimum: the 305 head and the S/R (Stock Replacement) head. The small-chambered 305 will go on the 355ci engine, and the S/R will go on the 406. Both of these heads are intended as stock replacements for street use. As basically detuned versions of the more expensive racing heads, and with the same rugged features, Dart heads have all of the bolt holes for accessories such as power steering and alternator.

All Dart cast-iron heads feature added material in critical areas to make them more durable and resistant to abuse. The decks are 0.400in thick, compared to about 0.250in on a stock Chevy head. This makes the Darts very crack resistant. We have seen these heads come off of

An air driver drives them in the rest of the way. The press fit between the new guide and the reamed-out guide holds them in place.

The excess guide is trimmed away flush with the original guide to give a stock appearance.

stock car engines that have been extremely hot. Stock heads would have cracked beyond repair; the Dart heads were barely warped. Their improved water passages and extra thickness facilitate better cooling and help eliminate hot spots which can cause detonation.

Exhaust seats are steel inserts designed to withstand unleaded fuel or the abuse of alcohol in a race car. These seats are machinable, meaning that they may be ground without danger of going beyond the hardening as can happen with induction hardened seats.

Screw-in studs are standard. The valve cover rails are machined to provide a positive fit of the valve cover gaskets. This helps eliminate potential leaks. Stock heads are notorious for valve cover leaks.

The one element lacking in the Dart heads is hardened or bronze guides. The guides do seem to be durable, but if you are going to run alcohol, bronze is a *must.*

The S/R (Stock Replacement) head is indeed a direct replacement that can be simply bolted on. We chose this head because it is economical and comes in an optional small-chamber design. We could have used the Dart Sportsman head, which will make more power. But it costs more money, and we set out to keep cost at a minimum.

We cc'd the heads and the chamber checked out very close to 67 cubic centimeters (cc) at 66.5cc. We also cc'd the intake ports—they are 168cc. This is somewhat larger than most stock Chevy heads, but not so large as to hinder low-speed flow velocity.

We chose to buy the heads bare and install our own valves and hardware. This way, we were able to cut the seats for 2.02in intake and 1.60in exhaust valves. The larger valves should help flow, thus allowing the engine to breathe better. Compare the price of fully assembled heads to the cost of buying the valves, springs, retainers, and seals separately. The cost may come out to be nearly the same, depending upon what you pay for the hardware.

Even if it is a little more money, we suggest buying the heads bare, buying the additional hardware, and having the machine shop assemble the heads for you. They can check the installed height of the springs and the spring pressure, and make sure the valves are seating correctly. Of course, if you use the normal-size valves, 1.94 and 1.50in, you will not have to pay for any machine work—the valves can just be installed. Our heads needed surfacing, even though they were brand new. Have the shop check the surface on yours.

We found that even though the S/R is made for the 350 Chevy engine, it fits much better on the 400 engine. The reason is because the combustion chamber in the head is larger than the 4.000in bore of the 350 block. Though this is quite common on stock Chevrolet engines, it is not the ideal setup for power and efficiency. When the combustion chamber is larger than the bore, it provides a "dead space" for fuel to hide and not be burned during combustion. The S/R heads fit quite well on the 406, though.

Torquer Cylinder Heads

World Products, in response to the high-performance use of their S/R Series cylinder head, now makes an S/R Torquer head. This is exactly the same head as the S/R with all of the same features, but it contains 2.02in intake valves and 1.60in exhaust valves. It, too, comes in 76cc and 67cc versions. This head sells for the same money as the regular S/R head, so buy it and save the expense of machine work for the larger valves.

The 305 Dart heads come with seats cut for 1.94in intake and 1.50in exhaust valves. We retained these sizes, since going to anything larger would not be practical. They are larger than the stock 305 intake valve size anyway, and should provide enough room for flow.

The 305 is intended for use on the 305ci engine—at least, this is what the World Products catalog indicates. We decided they would make power on the 355 because of the small combustion chambers. With the dished, stock replacement pistons, the engine has 9.3:1 compression. The

After the work is finished on top of the head, it is flipped over on the seat and guide machine. The bottoms of the guides are trimmed flush with the original casting just as they were on top. A 30- to 45-degree chamfer is put on the chamber end of the exhaust guides. This helps to dissipate heat and prevent the valves from sticking.

If the intake guides need to be knurled or the new exhaust guides need to be sized for the valve stems, this is done next. Once all the guide work is complete, seat work may be done. Here, the exhaust seats are being counterbored for seat ring installation.

chambers fit the 4.000in bore well, too. We scribed the heads to match the bore and unshrouded the valves out to this line. (More on how to do this in the section on porting.)

The heads received a minimum of porting. The sharp edges in the bowl area were blended and the short side radiuses were smoothed. The heads did need to have casting slag removed from the ports—they were rough and it is possible that some of the looser stuff could come loose and be inhaled into the engine. Scraping the ports with a screwdriver will knock the slag loose.

Cylinder Head Porting

In order to do elaborate cylinder head porting, a flow bench is necessary to check the results from changes made in the size and shape of the ports. A flow bench, however, is beyond the financial means of the average backyard engine builder. But there are some minor head modifications you can do in your home workshop, without a flowbench, that will make notable increases in horsepower.

Porting Tools

In order to do your own cylinder head porting you will need some special tools. An electric die grinder is the heart of the head porter's equipment; the rest of the tools are an extension of it. Be sure to get a die grinder with a collet rather than a chuck. Also, the collet has to be 1/4in. There are 1/8in die grinders on the market and they work fine for detailed hobby work, but for grinding away large amounts of cast iron they will not hold up.

Another important feature to look for is the die grinder's shape; it should taper down to the collet rather than end abruptly. This allows you to reach deeper into the ports. An assortment of carbide cutters with 1/4in shanks ranging from 3/8 to 1/2in in diameter are need-

The seat rings are driven in with a hammer and driver, and are retained by a press fit of about 0.0045–0.0050in. With the completion of this step, the old heads are ready for seat grinding (or cutting) and assembly. They are now able to withstand the rigors of unleaded fuel.

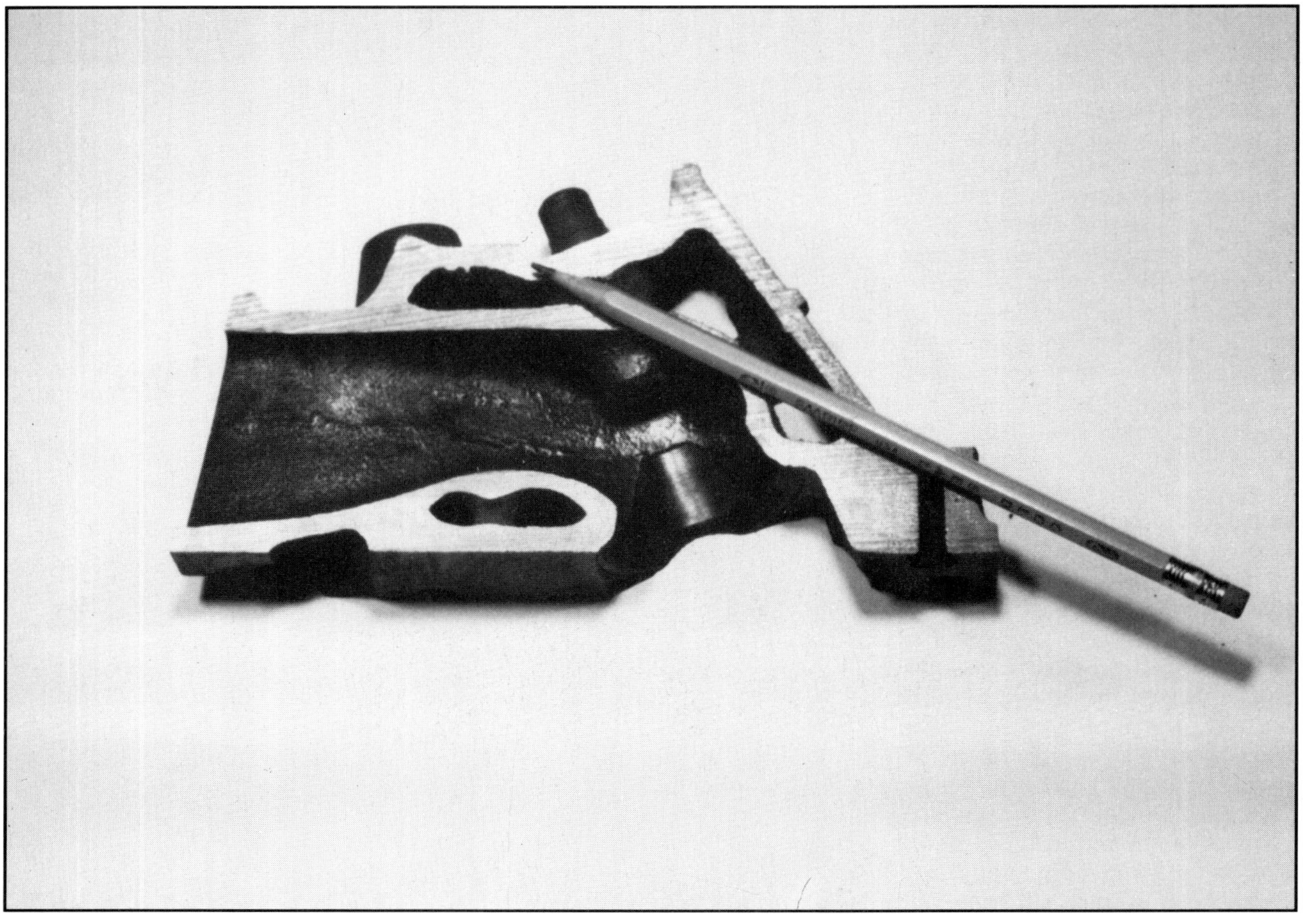

Even after all the machine work to update the old double-hump heads is completed, you still have a stock casting that is rather fragile in places. Note that the amount of material between the spring seat and the water passage is fairly thin. This is why it can be risky to enlarge the spring seats on these heads.

ed for rough grinding. You may use stones instead of carbide if you prefer, but they do not cut as quickly.

To smooth the work of the cutters you will need a selection of sanding rolls varying in diameter from 1/2 to 3/4in. Some 1/4in mandrels are needed to mount the rolls.

For the final polishing, 80 or 120 grit emery paper can be used. It is cut into strips and mounted on a flap wheel mandrel. The die grinder has too much rpm and will cause the emery cloth to shred quickly. To prevent this, get a rheostat control for the grinder, or use an air grinder.

Safety glasses and a face mask are essential items for this work. The glasses offer protection from grit and cast iron, as well as from a broken tool. The mask protects your throat and lungs from grinding dust and the adhesives that are used in the sanding rolls and emery cloth. Just as racers protect themselves on the track, protect yourself in the shop.

Grinding the Bowls

Now that you have all of this porting equipment laid out on the workbench, aren't you just itching to unleash some horsepower out of those heads? Where do you begin?

The first area to attack is the bowl region, just beneath the valve. You will see lines and ridges left from the factory casting and machining. The idea is to smooth and blend the lines and roughness, which act as restrictions to flow, to give the heads more capacity.

First, we will do the intake ports. To know just how much to grind out of the bowl area, measure the head of the intake valve and multiply this number by 0.9. If the valve is 1.94in in diameter, multiplying by 0.9 gives 1.75in. This is the desired diameter of the bowl, just beneath the intake valve seat.

You may think that bigger is better, but without a flowbench to test the results, bigger may actually *lessen* flow. Stick with multiplying valve head diameter by 0.9, and do not go beyond this figure—you cannot go wrong. Now set your dial calipers at

The deck thickness on the stock Chevy head is only about 0.250in. All you have to do is overheat them and they warp considerably and will even crack.

1.75in and use them to set a pair of dividers. With the dividers, you can check the progress of your grinding and you will know when you have reached the desired 1.75in diameter.

Begin with a carbide cutter in the die grinder, working the grinder in a circular motion following the contour of the bowl, smoothing away the roughness and blending the machining lines. As you reach the end of a stroke, lift the grinder slightly to avoid a dished-in spot at the point where the grinder pauses. Go slowly. It will take time to acquire a feel for this, but be patient and persistent. You may be more comfortable practicing first on an old junk head to get the feel of things and see what kind of results you can achieve. Then, when you feel confident, go to work on the good set.

Continue smoothly blending the bowl area, periodically stopping to check the progress with the dividers. When you reach the area where the floor of the port curves toward the seat (the short side radius), smooth the radius and blend it into the floor, but do not lower the floor of the port. All you want to do here is smooth the sharp edge off of the radius to provide a smooth path for the air-fuel mixture.

The material should be removed equally all the way around the bowl; do not cut heavily on one side, thus distorting the original shape. Simply smooth and blend the bowl.

When you have reached the desired diameter (1.75in with a 1.94in intake valve), you are done with the intake bowl. Just repeat the process on the other seven so they match as closely as possible.

After completing the intakes, move on to the exhaust side. Again, measure the head of the valve. But, this time, instead of multiplying the valve diameter by 0.9, multiply by 0.85 to determine the bowl size. If the exhaust valve measures 1.50in, the bowl should be 1.38in. The reason for the smaller multiplier on the exhaust side is to create a venturi effect in the exhaust port to help exhaust gas velocity. Again, as you did on the intake bowls, smooth and blend the

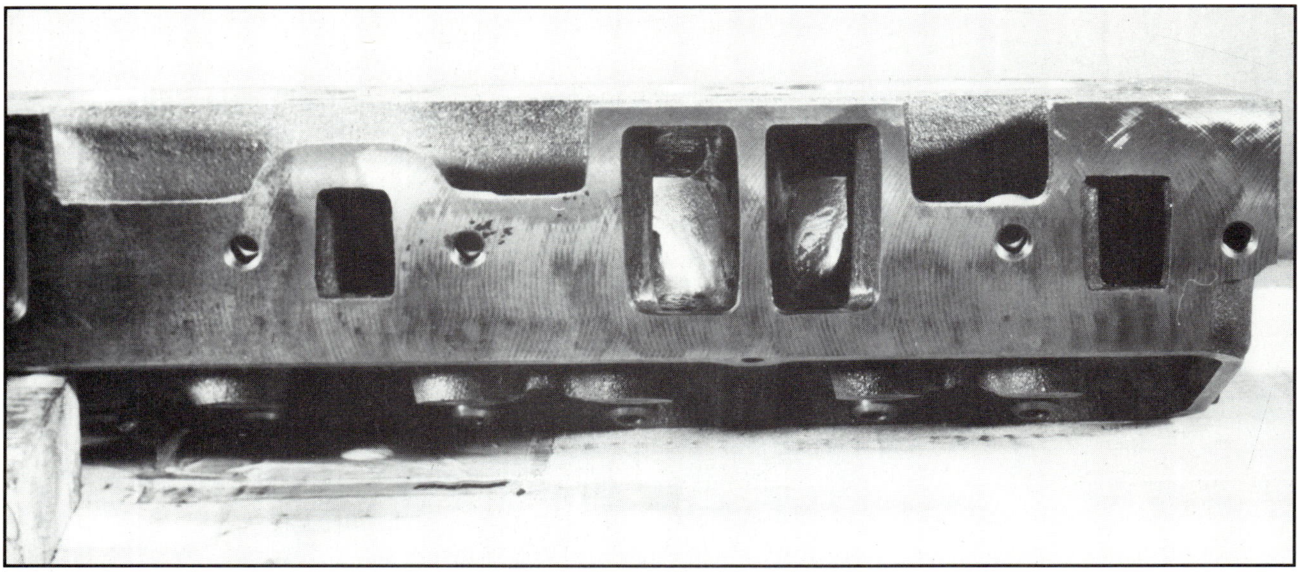

The Dart heads come a little rough in the ports. We used a die grinder in these intakes to knock off the big chunks. Look closely and you can see the marks; no real porting work here, just slag removal. At least take a screwdriver and scrape the ports to get rid of the loose stuff.

We found that our Dart heads did need surfacing. Seems like they should be perfectly flat, but this is a common problem with many new castings, probably due to production-line surfacing techniques. If you buy a set of new heads all assembled and ready to go, it would be wise to take them to a machine shop and have the surfaces checked. Be sure the shop uses a rotary broach as is pictured here, or any surfacer that uses carbide inserts or a stone. Stay away from belt sanders for head surfacing. Many times they do not leave a truly flat surface, and after shelling out your money you will not have anything better than you started out with.

area, following the contour of the bowl. Periodically check with the dividers until you reach the proper size.

One advantage to installing larger valves is that it allows you to enlarge the bowls more. Larger head diameters multiplied by the constants 0.9 and 0.85 yield bigger numbers. Being able to remove more material allows you to blend and smooth the bowls even more, as well as provide larger openings for flow.

After the rough porting is done, use the sandpaper rolls to smooth away the rough surfaces left by the carbide tools. If you are smooth and steady with the grinder you will have less sanding to do. For an even smoother finish, use the emery cloth in the flap mandrels for a final polishing job.

Smoothing the Ports

After finishing the bowl areas, move on to the intake ports where they meet the intake manifold. The goal here is not to change the port, but to square it up.

First, you need to make a pattern to follow. Do this by placing an intake gasket on the head, holding it in place with manifold bolts, then scribing its pattern on the head surface. If you coat the intake side of the head with machinist's bluing or a light coat of paint, it will make the scribe marks show up better. Be sure to use a stock gasket—a racing gasket is too big, and you will get into trouble trying to match the ports to it.

Just as you did in the bowl area, use a carbide cutter in the die grinder for the initial work. Then smooth with sanding rolls.

In the upper corner of each intake port is a lump of casting for the valve cover bolt. You may break into the bolt hole when squaring the ports. There is no need for panic if you do so; simply use valve cover studs instead of bolts. The stud will still fill the holes when the cover is removed, thus preventing dirt from falling into the ports.

All valves, even new ones, should be faced on a valve grinder. The valves should be faced with a 45-degree angle so that it perfectly matches the 45-degree seat angle. Gone are the days of grinding an interference angle, that is, putting a 44 1/2 degree angle on the valve. With new valve and seat equipment, valves will seal instantly. Also, with a narrow contact point created by an interference angle, the heat cannot dissipate quickly, and damage to the valve or valve sticking may result.

Just as in the bowl areas, there may be casting seams in the ports. However, some of these seams you may not be able to remove entirely. To do so would entail excessive grinding in the seam area, which in turn would cause a dip in the port. This dip, or low spot, will interrupt flow. So, if there are some irregularities left in the port after the

We enlarged the seats on the Dart S/R head to accept 2.02in intake valves and 1.60in exhaust valves. This intake is being cut in the Serdi seat and guide machine. All three angles are cut at once, thus ensuring precision.

squaring up, don't worry about them.

Another potential problem is bellmouthing the end of the port. This is quite easy to do if you are not careful. It results from grinding the end of the port out to match the gasket, but not carrying the work back into the port. Bellmouthing leads to a dramatic increase in size where the port meets the manifold and causes turbulence in the air-fuel mixture. Be sure to carry the work back into the port to prevent bellmouthing. The port has to be the same size or larger than the manifold port. Otherwise, it will be like erecting a brick wall for the mixture to try to get over, thus creating turbulence.

Intake ports on small-block Chevy heads have to bend around the pushrods. This bend should be straightened as much as is practical. A lot depends upon how brave you are—grind too much and you will break through the side of the port. There is no real need to worry, though. Epoxy will take care of any small holes if you get carried away. Blend the port area around the pushrod into the rest of the port to make as consistent a path as possible. Remember, drastic changes destroy flow, so make the changes in size and shape as gradual as possible.

After completing one port, use it as a pattern for the other seven. Again, you can use dividers for comparison purposes. Measure the first port, then measure the same spot in the other ports to maintain consistent sizing. For instance, you can first measure the area just inside the port, then the other seven back that far. Next, measure a little deeper in the first port and work your way back until you have all eight intakes ported equally. Working in this manner will help maintain equal flow to all cylinders. And all cylinders working equally means more horsepower!

Now for the exhaust ports:

Exhaust seats are cut the same way as intakes, with a carbide bit containing three angles. The dial indicator attached to the spindle is for equalizing seat depths. The operator cuts the first seat, notes the reading on the indicator, and cuts all the other seats the same.

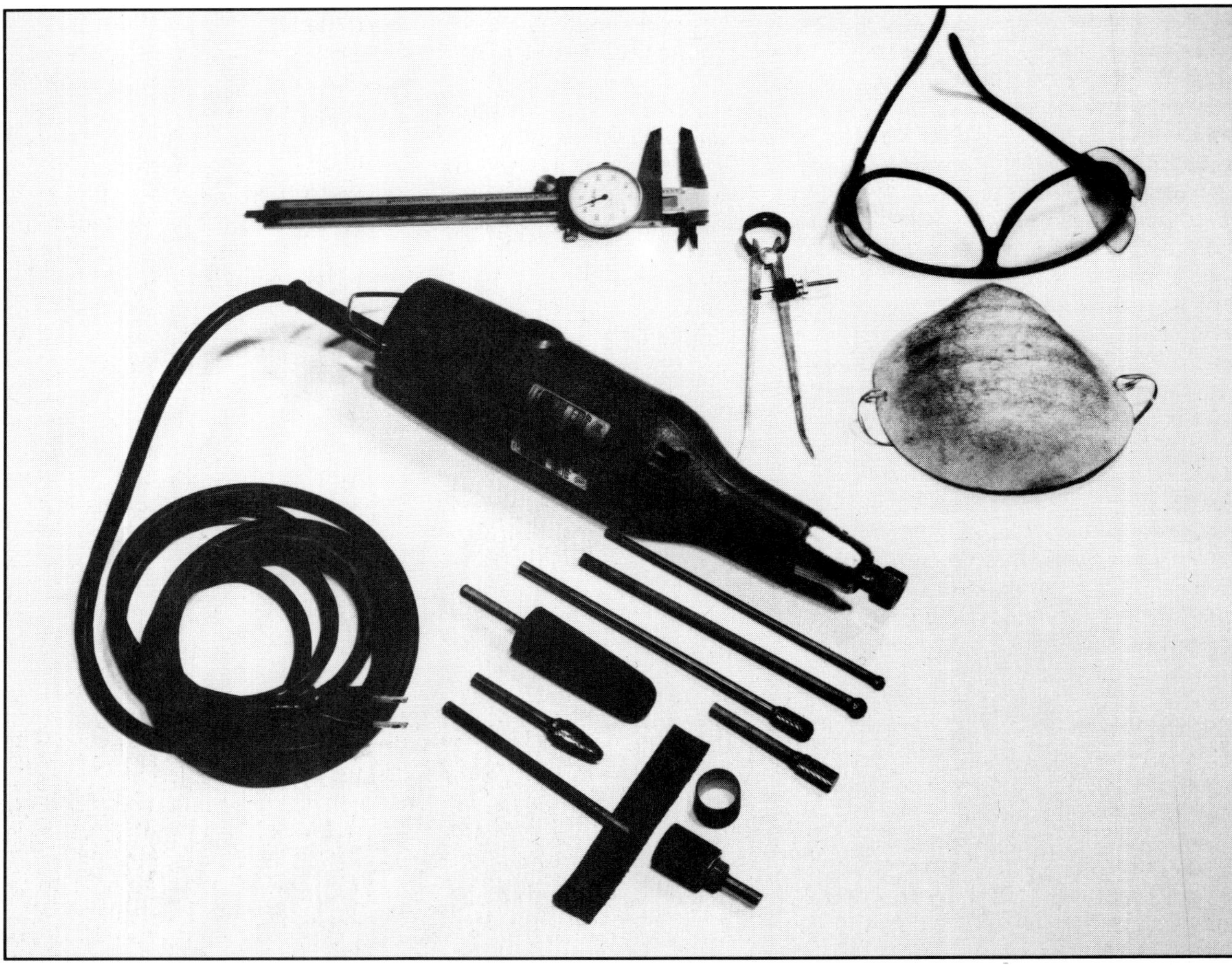

Necessary tools for head porting. The electric die grinder is the heart of the equipment, with the other tools being an extension of it. Note how the grinder tapers down to the collet, thus allowing it to go deep into the ports. Dial calipers and dividers are a must for accurate measuring. The mask and safety glasses are just as important as any of the other tools; they keep the dust out of your eyes and lungs. The glasses also protect from flying pieces in case a bit or stone shatters. Do not even *think* about doing this work without proper protection!

leave them alone. The only thing you need to do is grind away any unusually large lumps of casting if any exist. The exhaust ports flow well to begin with on Chevy heads, and without a flowbench you can get into trouble quickly. It is better to leave well enough alone rather than taking a chance on making a good thing bad.

Shaping the Combustion Chambers

Any rough spots in the combustion chambers should be ground away. A rough piece of casting can become a hot spot, thus causing preignition. The valves can be unshrouded to aid flow. The combustion chambers, at the point where they come closest to the valves, pinch the flow. If the chamber is laid back, it will allow for freer flow.

To determine just how far you can lay back the chamber, spray the deck of the head with machinist's bluing or a light coat of paint. Bolt the head to the bare block and reach up through the cylinders from the bottom and scribe a line on the head using the cylinders as a pattern. Remove the head and grind the chamber, at the point where it is closest to the valves, back to the scribed line. Do not go beyond!

You should not use a gasket for a scribing pattern because the gasket is bigger than the bore. Your pattern will be too big and you will lay the chamber back too far. Remember, it should not be bigger than the cylinder bore.

When you unshroud or polish chambers, be sure to use an old intake and exhaust valve to protect the seats.

All of this grinding work should be done before the heads

have had a valve job. The reason is because it is easy to accidentally get into a seat with the grinder, thus ruining its sealing ability. The valve job would then have to be redone, increasing your expense. Even though the seats have not been ground, you still need to be careful because if you damage a seat too deeply, it will have to be replaced. If you port your Dart heads, the seats will already be done. You will just have to be *very* careful.

The porting procedure described may be used on either stock Chevrolet heads or the Dart heads. You may wonder just how much power improvement you can gain from this work. The answer is somewhat generalized; it is hard to pinpoint specific figures. You can expect a gain of 10–40 percent, or about 15–60 cubic feet per minute (cfm) per port of flow at peak valve lift.

The reason for the wide range of flow increase is due to the differences in heads. Some castings flow better than others in stock form, and even castings with the same part numbers (which should be identical) will flow differently due to foundry and machining processes.

The better a head is to begin with, the less effect porting will have on it. Thus, you will not gain as much from a better-quality head as you will from a poorer-quality type simply because there is less room for improvement. The fact remains, though, if you do the work and do it properly, there will be a gain in flow and a resulting gain in horsepower.

Valves

For both of our engines, we used economical replacement valves from E-Loy.

Don't spend money needlessly on swirl-polished valves, stainless valves, or any other gimmicks. Stainless withstand higher heat, but are not necessary for the applications here.

The important factor to consider in buying valves for your engine is the stems—chrome stems are far superior. The reason is because they are smooth. A stem that is not chrome plated is rough and will destroy the guides. As the valve moves up and down through the guide, it has the same effect as running a tiny file through it. The guide will be worn away. The smoothness of the chrome-plated stem

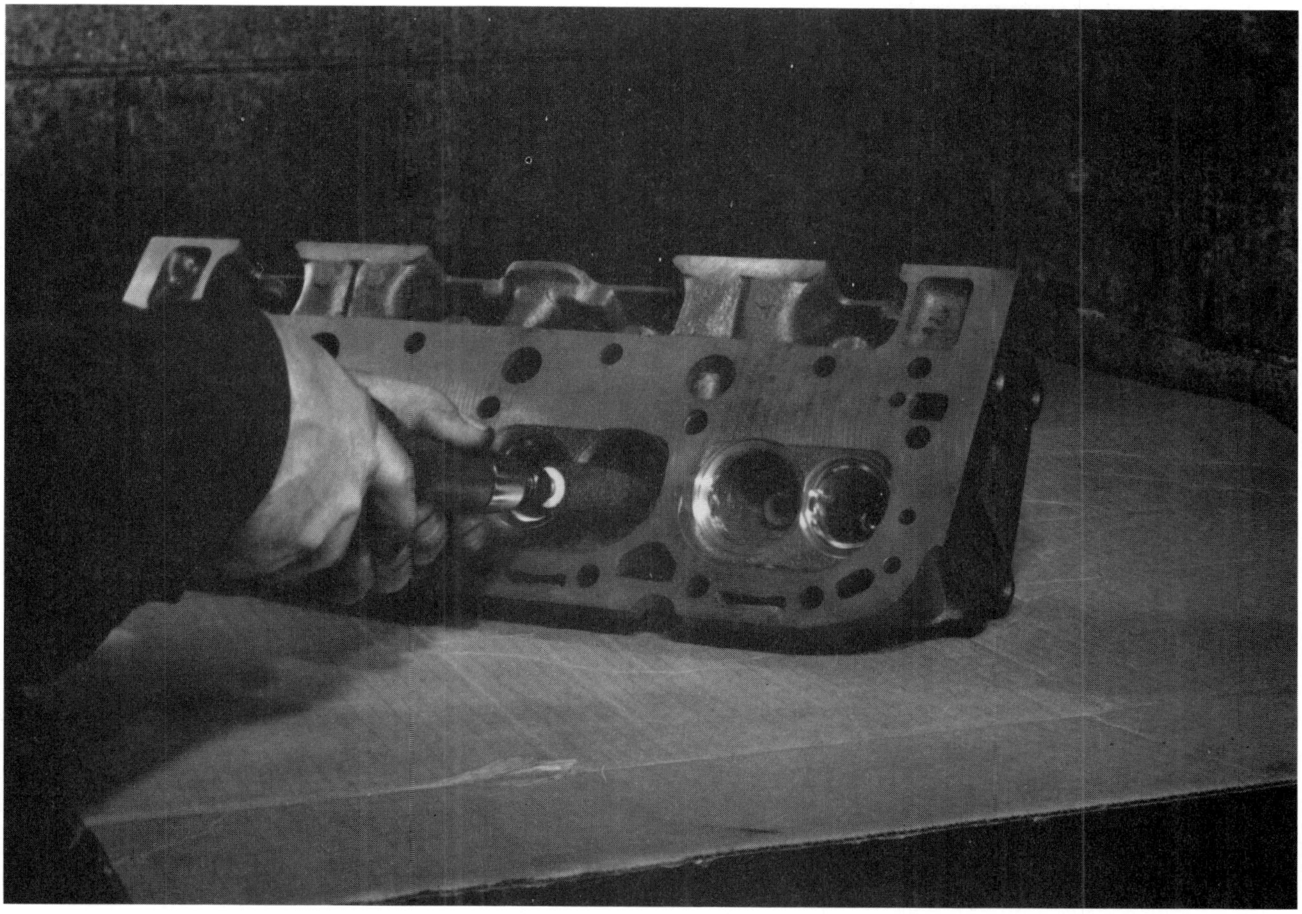

Porting is dirty work, so do it in an area away from assembly. Work the stone or carbide cutter in a circular motion, back and forth, following the contour of the bowl. Smooth and blend until the correct diameter is reached.

Look closely and you can see the scribed lines around the intake ports made by using a gasket as a template. Open the ports to match them. Blend the work down into the port so that there is not a sudden enlargement, or bellmouth, at the end of the port.

The object of the work in the bowl area is to blend and smooth the lines and roughness left by factory machining.

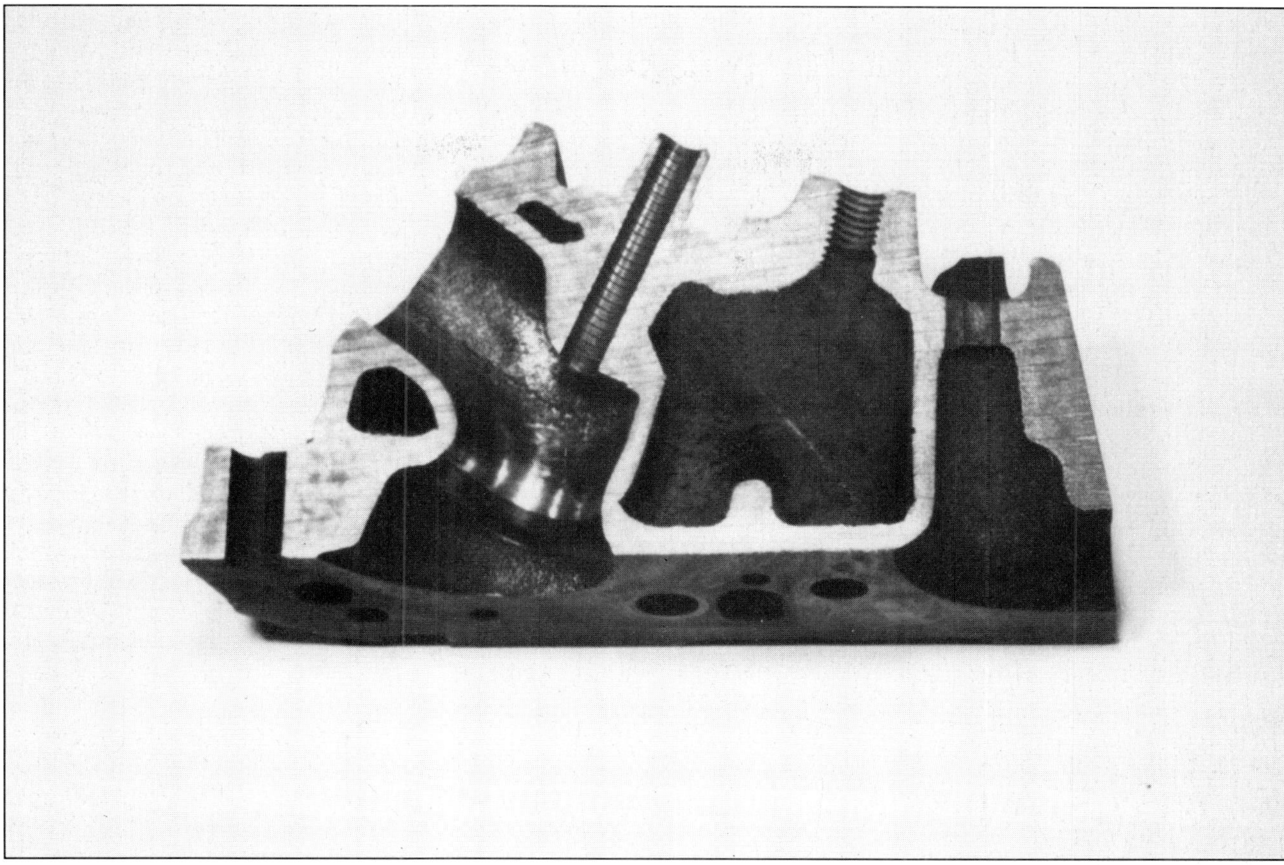

The exhaust bowl after porting. Notice how much smoother it is now. Also note the porting depth—just far enough to get rid of the factory lines. The short side radius, the area where the bottom of the port turns down toward the seat, should be smoothed. There is an abrupt change of direction there, so just smooth it to form an even-flowing path.

means far less wear to the guide and therefore a longer lasting guide and valve. The E-Loy valves have chrome stems.

Valve Springs and Retainers

Big valve springs often come to mind when high-performance engines are discussed. Sometimes big springs are a necessity, but on these engines they are not needed. You are far better off using springs that are nearly stock in diameter and have a reasonable amount of pressure than you are using large-diameter springs with loads of pressure. There are two reasons for this. First is cost. Springs that fit the stock spring seat simply are less expensive.

We used LT-1 springs from Elgin on the 305 heads. These springs are good for valve lift of up to 0.500in and therefore are compatible with the 903-P Elgin hydraulic camshaft. They are used with the stock retainers and require no spring seat modification. Money is saved on machine work and the springs themselves are inexpensive, costing only $25 a set.

The springs for the 406, which has a higher-lift cam, are also inexpensive. They are from Competition Products, part number 327, and cost $59.95 plus shipping. These double springs can handle valve lift up to 0.600in. Though they do not fit the stock retainers perfectly, the fit is close enough not to present a problem. The spring seat does have to be cut down around the base of the guide so that the inner spring will seat properly. This is an operation you can do yourself with a drill and a modified spring seat cutter.

To make you own cutter, order one from most any of the cam manufacturers. Buy the cutter with the smallest outside diameter possible, since you do not want to enlarge the seat anyway. With a bench grinder, grind off the outer tips of the carbide inserts so you do not enlarge the seat. Leave intact the inner edges of the insert that cut the area around the guide. Put the tool in the drill, and cut down the inner seat area so that it is flush with the rest of the seat. Now the inner spring of the Competition Products 327 springs will fit properly.

Stock retainers are cheap and strong. Aftermarket retainers can be quite expensive and oftentimes require special locks to be used with them, which means even *more* money to spend as well as the additional machine work. By utilizing the LT-1 springs and Competition Products springs, we saved money on the rest of the hardware and the machine work.

Screw-In Studs

The Dart heads come with the stud bosses drilled and tapped for screw-in studs. With the LT-1 springs this is not a necessity because there shouldn't be enough spring pressure to pull out pressed-in studs, as each set of stock Chevy heads has. But, the additional insurance is nice and if you want to add bigger springs such as we did on the 406 engine, you're all set.

Guide Plates

Guide plates are somewhat of a myth. It is just assumed that if you have a high-performance engine, guide plates are a requirement. Not true. The small-block Chevy cylinder heads, including the Dart heads, come with pushrod slots in them. These slots hold the pushrods in proper alignment and they do their job quite well.

Only under two circumstances are guide plates needed. First is if you deviate from the 1.5:1 ratio of the stock rocker arms. On the 406ci engine we are using rockers with a 1.6:1 ratio to gain some valve lift. In this case, the rocker-arm socket in which the pushrod locates is closer to the fulcrum point, or the stud, than stock. This pulls the pushrod forward so far that the slot in the head is not long enough to properly guide it. There are two options here. With a special tool called a Louis tool, which can be purchased at any speed shop, the slot can be enlarged with a drill and appropriate size bit. Or, you can open up the slots with a large drill bit or a high-speed grinder and then install guide plates.

A second situation in which you would need guide plates would be if the heads already had the stock pushrod slots opened up. Then you would have to install the plates, since the opened slots would not direct the pushrods.

The intake port after porting. Notice how it, too, like the exhaust port, has the lines blended and smoothed. The short side radius here, too, needs to be blended to provide an even path for the mixture.

To scribe a head for valve unshrouding, this is the way to do it. Blue the deck surface, bolt it to the block, without a gasket, and reach up underneath and scribe the cylinder bore onto the head. You can see that the intake side of the combustion chamber is smaller than the cylinder bore. (The exhaust does not show up here.)

This is what the head will look like after you remove it from scribing.

If neither of these conditions exists, do not waste your money on guide plates. Remember, guide plates are needed only if you change rocker-arm ratio and do not use a Louis tool to elongate the stock slots, or if the heads have had the slots opened up.

Cylinder Head Assembly

If you order Dart heads fully assembled, you won't have to worry about cylinder head assembly. But we recommend you buy the heads bare and have them assembled or do it yourself.

To do your own assembly will require some special tools. A valve spring compressor is a must—there is no other way to compress the springs enough to get the retainers on. You also will need a method to check spring installed height and pressure. A good valve spring tester is quite expensive and does not warrant the cost unless you plan to do a lot of head work. Portable spring testers can be bought that work in a vise, and the cost is far less than the professional benchtop model. A valve spring height micrometer is the best way to measure spring height, but dial calipers or a 6in steel machinist's scale will work; however, the accuracy will not be as great.

Clean each guide with solvent, then place a drop of oil in each and smear it around thoroughly with a small rifle-cleaning brush. Lube the valve seat, too, with a light coat of oil. Liberally oil the valve stems and place them in their respective guides. Work them up and down a few times to make sure they slide freely. If they don't, find out what is wrong. The valve stem may have a nick or burr on it, or the end of the guide may be burred. Correct the problem before proceeding. If you have to deburr a guide, be sure and clean it thoroughly and oil it again before assembly.

Once the valves are in place, put a retainer and keepers on an intake and exhaust valve. Leave the springs off. Check the distance from the bottom of the retainer to the spring seat. This is the installed height of the spring. If you have a means to measure spring pressure, measure the spring at this height. This is the seat pressure. If you lack the means to measure the pressure, take the spring and retainer to the machine shop along with your installed height figure and have them check it for you. Now, subtract the valve lift from the installed height to find out what the open pressure will be.

For our hydraulic cam in the 355 with a lift of 0.447in, the 0.447 is subtracted from the installed height figure. Test the spring pressure at this height. This gives you the open pressure. Also, the spring should be checked at this time to make sure there is enough space between the coils at full valve lift. There has to be at least 0.030in clearance. There was no problem with the cams and the springs we used. This should always be checked, though, whenever either the cam or the springs or both are not stock.

The LT-1 springs on the 355 have 100lb seat pressure, and the 327 springs have 120lb. Open pressure for both should fall in the range of 280–300lb.

After determining the correct installed height, check each valve and retainer combination and make sure they are all equal. Use shims as needed. When assembling the springs, place the retainer on the spring, compress the assembly with the spring compressor, slip the O-ring seal over the stem and squarely in place in the bottom groove, locate the two keepers in the top groove, and gently release the compressor. Do this fifteen more times, and the heads are assembled.

With a soft hammer, rap each valve stem to make sure the keepers are in place. If they are not, they will pop loose. Repeat the assembly on any valves where the keepers come loose. Now is the time to discover a mistake, not when it is back on the engine.

Above right
We put the head back in the Serdi for valve unshrouding. Using a radius cutter, we laid back the chamber to the scribe marks on both intake and exhaust. The smooth, shiny areas on the chamber wall are the areas that were cut away. The chamber walls pinch the flow next to the valves, so by laying this area back, the engine breathes a little better. It could be done by hand with the die grinder, but this method is much faster and produces nicer looking, more consistent results.

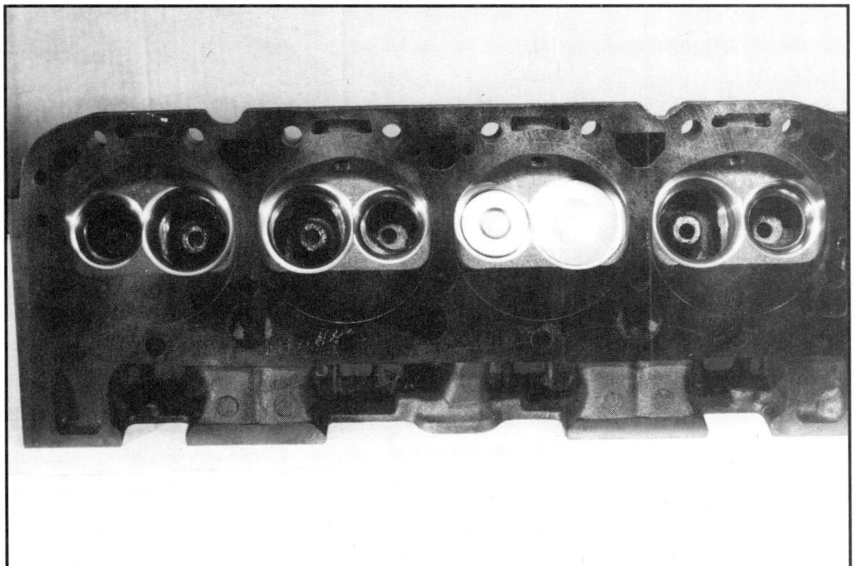

Right
The results of the unshrouding: There is much more room between the valves and chamber walls.

Our cylinder head hardware—nothing exotic! Economy valves, stock retainers, stock keepers, stock O-ring seals and oil shields. The springs are LT-1, which is all you need for the mild hydraulic cam in the 355. The 406 used a set of double springs that fit with the stock retainers. Pretty simple, very inexpensive, and very low maintenance.

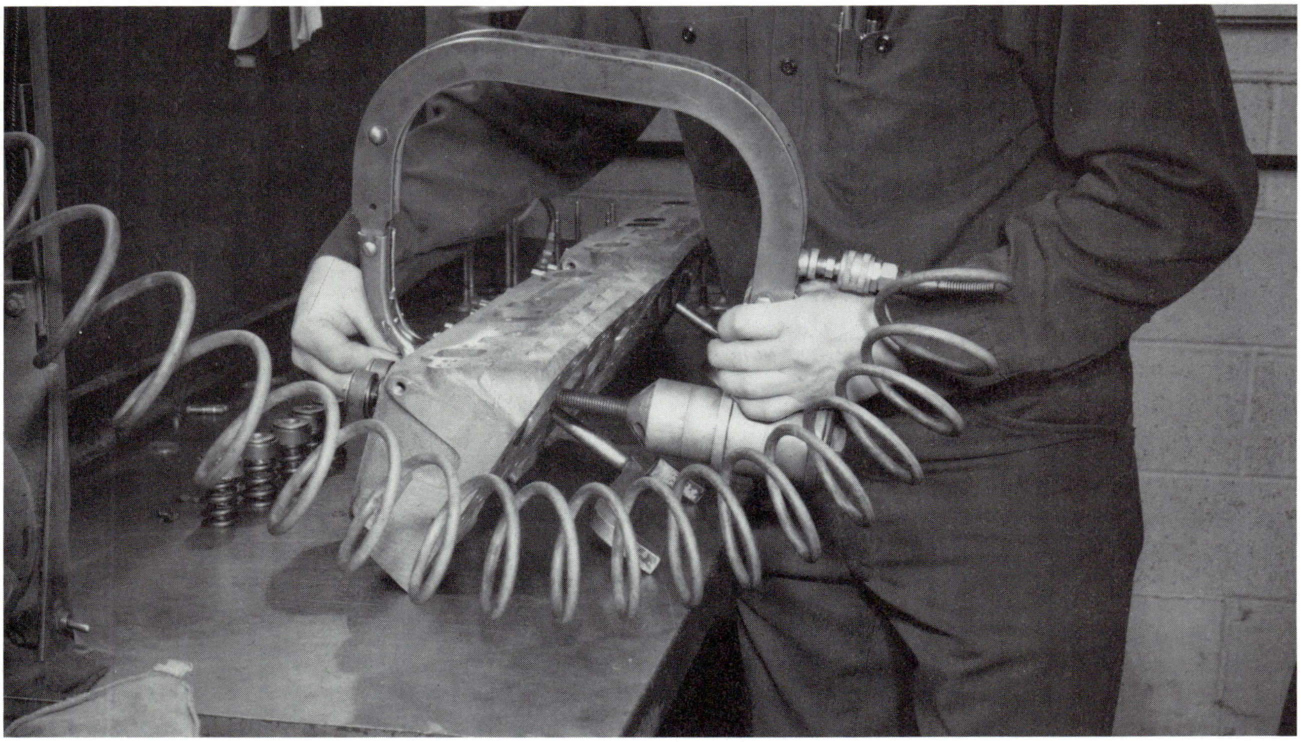

This air-operated spring compressor makes the job easy. A manual one will do the job, it just takes a little more muscle. Coat the inside of the guides and the valve stems with oil before assembly. Place the oil shields and retainers on the springs and compress them. The oil seal comes next, and goes into the bottom groove. Put the keepers in place in the top groove and slowly release the compressor.

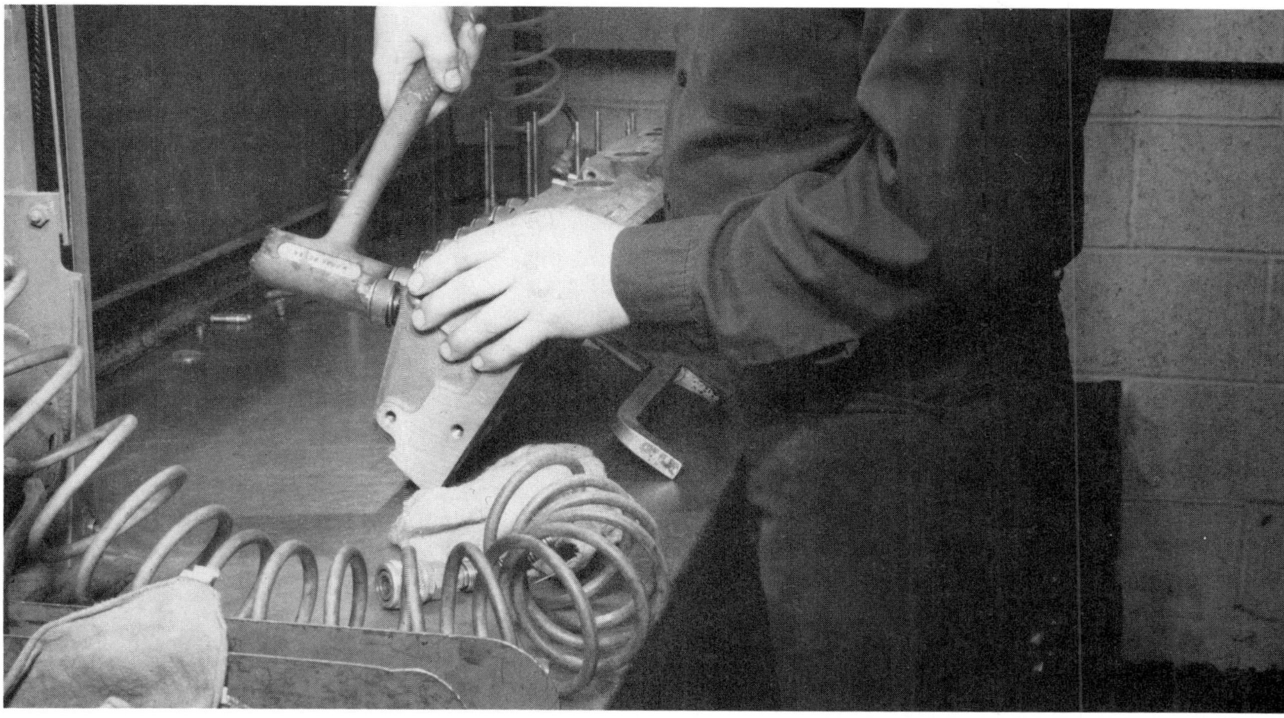

Rap the tops of the valve stems with a soft hammer to make sure the keepers are solidly in their grooves. If not, they will pop loose. Better to find out now than on the engine.

To test and see if the oil seals are seated properly, use a hand vacuum pump and the special attachment that fits over the valve retainer. If vacuum is held, the seals are doing their job. If not, rap the valve stem again to help seat the seal. Check again. If it still doesn't hold vacuum, the seal may be damaged or out of its groove. Disassemble, recheck, and try it again.

On stock Chevy heads, sealer must be used on screw-in studs. With World Products Dart heads, no sealer is required unless the studs go into an intake port, because the studs do not go into water. Torque to 45lb-ft.

If you are going to use 1.6:1 rockers, the pushrod slots in the heads must be elongated. One way to do this is with a hand drill and a Louis tool. The tool fits into the stock slot and positions the drill bit to properly elongate it. The other option is to drill out the slots and install guide plates.

Cam Degreeing and Valvetrain Geometry

6

Degreeing in the camshaft is not necessary for the engine to run. Many engines are run successfully and competitively by simply installing the cam right out of the box. Tolerance "stack-up" may cause the valve timing to vary 2–3 degrees, however. Each company has a tolerance range for their work. A particular piece may fall to the high or low side of this tolerance.

Cam timing depends upon the timing gear dowel in the end of the cam, the keyway in the crankshaft gear, and the keyway in the crank. If all of these items are to one side of their tolerance, they may all stack up to throw off the timing. Degreeing the cam will let you find out if this is the case in your engine.

To degree a cam you will need some special tools: a degree wheel, a dial indicator, a fixture to mount the indicator to the crank and allow you to rotate the crank, and a couple of custom lifters. All of this will add up to around $100.

Cam Degreeing Procedure

There is more than one way to degree a camshaft. The end result is the same, however: you are determining the motion of the valves in relation to the motion of the pistons.

The method we will use here is

The first step to degree a cam is to find top dead center on the number one piston. You need a degree wheel, a pointer, and a TDC stop. Notice the stop is spaced up off the deck of the block with some thick washers. The piston has a dome, so the TDC stop had to sit up high enough to clear it. A piece of dipstick makes a good pointer. Rotate the engine until the piston comes up against the stop. Note the degree wheel reading.

Rotate the engine the opposite direction until the piston comes up against the stop. Note this reading.

the easiest we know of—it shows you the relation of the cam's lobe centerline to the position of the piston. Therefore, it is called the lobe centerline method of camshaft degreeing.

To begin, insert the piston in the number 1 bore. No need to install rings for this. Also, just snug the rod bolt nuts; don't torque them down. Remember to install the bearing and apply a coat of oil to it.

Affix the degree wheel to the end of the crankshaft. A pointer will be needed. We use an old dipstick cut to proper length, with a grounded point on the end. It can be fastened easily to the block with a bolt and a large, flat washer in a water-pump bolt hole.

The first step is to determine top dead center (TDC). A piston stop is needed for this. You can buy one, or make your own, like we did. To make a piston stop you need a 1/4in piece of steel plate large enough to bolt to the block and cover a piston. Use a head gasket as a template and drill three holes that line up with the three head bolts. In the center of the plate, drill another hole for a bolt that will act as an adjustable stop. Bolt the plate to the head with 7/16–14 bolts about 1/2–3/4in long.

Rotate the engine until the piston comes up against the stop. Go slowly so as not to damage the piston. Note the degree on the degree wheel. Then rotate the engine in the opposite direction until the piston comes up against the stop once again. Note the degree wheel reading again. TDC is exactly halfway between the two readings. Mark this point on the degree wheel. Remove the piston stop.

Now you need to find the point of maximum lift on the intake lobe on number 1 cylinder. Place the special lifter in the appropriate bore. Secure the dial indicator to the deck surface of the block, and set the indicator point on the lifter extension. Rotate the engine until the lifter reaches its highest point of travel. Continue to rotate until the lifter comes back down 0.050in. The reason for this is because when the lifter reaches its uppermost point of travel, it will dwell there for 2–3 degrees of rotation. Coming back down, 0.050in is an accurate spot to check the reading. Mark the

The point exactly between the two readings is TDC.

degree wheel when you reach this point.

Rotate the engine in the opposite direction and repeat the procedure. After the lifter reaches its highest point, continue to rotate until it comes back down 0.050in. Mark the degree wheel. Count the number of degrees between the two marks, and mark the point halfway between the two readings. This is the intake centerline.

The last step is to determine the relationship of the intake centerline to the piston at TDC. To do this, count the number of degrees between the TDC mark and the lobe centerline mark on the degree wheel. This distance, in degrees, is the lobe separation of the cam. If the cam is ground with a lobe separation of 110 degrees on the intake, as our 903-P cam is, and everything is in order, that is, the cam is accurate and the keyways are in spec, then the distance between the marks will be 110 degrees.

Repeat the procedure with the exhaust lobe, and determine the lobe separation for it.

To retard or advance the cam, simply move the cam so that the distance between the TDC mark and the lobe centerline mark is either more or less in degrees. It's that simple!

For example, with the 903-P camshaft, if you want to advance the cam 4 degrees, and currently it is straight up, move it so that the distance between piston TDC and lobe centerline is 106 degrees. If you wish to retard the cam, you would move it so that the distance between piston TDC and lobe centerline is 114 degrees.

To move the cam you can use a timing chain and gear set, such as we have used on our engines, that have three keyways in the crankshaft gear. You may also use offset bushings that fit into the camshaft gear dowel hole. The multiple keyway crankgear is the easiest method, since no modifications are needed.

Be aware of one important fact: Whenever you change cam timing, *check the valve-to-piston clearance*. Even if the valves do not touch the pistons with the cam installed straight up, they may touch if the cam position is changed. The valves will be opening in a different relationship to the pistons after the change—it

Now install the special lifter. This is just a stock lifter with an extension on it to reach the dial indicator and to provide a flat spot for accurate readings.

The indicator is held to the deck by a magnetic base. This is a fairly inexpensive setup and works well. The indicator does have to be held solidly for accurate readings.

Alternate Method of Cam Degreeing

If you do not have the cam-degreeing equipment mentioned, there is an alternate method. It is not quite as accurate and will not work on dual-pattern cams, but it will work for the cam we used in the 355ci engine.

First, you have to determine exact TDC. You can do this with the heads on if you make a TDC tool out of an old spark plug and a bolt. Remove the electrode and the porcelain parts of the plug, and tap it on the inside to accept a bolt. Remove the rocker arms from number 1 cylinder then simply screw the tool into the spark plug hole. Turn the bolt in far enough that the piston will hit it when it nears the top of its stroke. Rotate the engine one direction until it stops, and mark the balancer or degree wheel. Rotate the engine in the opposite direction until it stops again. Mark the balancer or degree wheel again. The point exactly midway between these two marks is true TDC.

Rotate the engine so that it is on exactly TDC and the cam is in the overlap position—this is when both lifters are on the bottoms of their cam lobes. Lay a straightedge across the tops of the lifters and, with a feeler gauge, measure any difference in their heights.

If the intake lifter is higher than the exhaust lifter, then the cam is advanced. If the exhaust lifter is higher, the cam is retarded. If both lifters are the same or within 0.005in, then the cam is straight up. As a general rule of thumb, 0.006in difference in the height of the lifters is equal to about 1 degree. For example, if the intake lifter was higher than the exhaust lifter by 0.024in, then the cam would be 4 degrees advanced: (0.024 divided by 0.006 equals 4).

To check valve-to-piston clearance, place modeling clay on top of the piston. Be sure to get it in the area underneath the valves—fill the valve reliefs.

may be just enough to cause problems.

Cam Timing and Power

Do not be under the illusion that changing the cam timing will make the engine produce more power. All it does is change the rpm range where the cam is most effective. Generally, advancing the cam will lower the power band and retarding it will raise the band.

If you need more power off the starting line or coming out of the corners, it may help to advance the cam. If you need more power at the far end of the strip or going down a long straightaway, then you may gain from retarding the cam. Remember, changing the cam timing is not a power secret, it just places the peak power into a different rpm range.

Valve-to-Piston Clearance

When using a camshaft with higher than stock lift, valve-to-piston clearance must be checked. If the valves strike the pistons when the engine turns over, the valves will bend and possibly the pistons will be damaged.

The TDC stop you can make from an old spark plug and a bolt. Very cheap, and it works. Just screw it into the spark plug hole.

The easiest way to check this clearance is by using the clay method. Purchase some modeling clay and lay it on top of each piston in the area where the valve reliefs are cut. Just as in the degreeing procedure, install the pistons without rings and snug the rod bolt nuts. Assemble an intake valve and exhaust valve in the head using special lightweight "test" springs. Oil the valves so the clay doesn't stick. Use an old head gasket of the same style in assembly, or use your good gasket, but don't torque down the bolts; just snug them. After the head is on, complete the assembly with pushrods and rocker arms.

Rotate the engine through at least two complete revolutions. Then remove the cylinder head and, with your dial calipers, measure the thickness of the clay at its *thinnest* point; that is, measure the clay where the valves come closest to the pistons. There *must* be at least 0.100in on the exhaust and 0.080in on the intakes. If there is less, then the valve reliefs in the pistons have to be deepened.

Iskendarian (Isky Racing Cams, Gardena, California) cams makes an excellent tool for deepening a piston's valve reliefs. You can cut the pistons right in the engine with the heads on. The valves are removed, and the valve guides are used to pilot the tool into posi-

Install an intake and exhaust valve using lightweight checking springs. Put the head on using an old gasket just like the one you will use. Hold the head on with two or three bolts just snugged down. Install the lifters, pushrods, and rocker arms, and set the lash. Turn the engine at least two complete revolutions.

Remove the head and check the clay at its thinnest point. This is where the valves came closest to the pistons. There has to be 0.080in clearance for intakes and 0.100in for exhaust. Any less than this and the valve reliefs in the pistons will have to be deepened.

tion. By using the guides to align the tool, the relief cutting is done in exactly the correct spot on the piston.

Rocker-Arm Geometry

If the rocker-arm geometry is not correct, the valve guides will wear out very quickly, and in extreme cases, in only a matter of a few minutes. The geometry has to be right for an efficient and long-lasting engine.

Many factors can affect rocker-arm geometry. If you have stock rockers and run a cam with over 0.500in lift, the geometry will be off; more than 0.500in lift, and aftermarket roller-tip rockers are a *must*.

Camshaft base circle can affect geometry as well. Our 1090-P cam in the 406 engine has a base circle 0.080in smaller than a stock cam, which means that the lifter sits lower in the bore when it rests on the base of the lobe. This, in turn, affects the geometry, which means you *may* have to make corrections such as longer or shorter pushrods.

In extreme situations, the rocker arm can actually move backwards across the valve stem as the valve is opening, pulling the valve against the guide. Under these circumstances, the guide would last for only a few minutes of hard running. Then the engine would use excessive oil and the valves would not seat correctly, resulting in significant power loss.

Look at the tips of the rockers and note where they are posi-

Rocker arms should be centered on the valve stems. The rocker on the far right is off center just a bit. Compare it to the others.

Establishing the Correct Oil Level

When you use an oil pan with more than stock capacity, retain the stock dipstick tube. When you fill the pan with oil refer to the full mark on the stick just as you would with a stock pan. This is important because this mark indicates the proper oil level in relation to the crankshaft.

If the aftermarket pan you use is advertised as being a 9qt pan, it may actually only take 8 1/2qt to bring the oil up to the proper level. Or, it may take 9 1/2qt. Just refer to the dipstick for the proper amount.

tioned on the valve stem tips. They should be well centered left to right and front to back. As a further check, blue the tip of the valve stem with machinist's bluing and install the rocker arm. Set the lash as closely as possible without disturbing the bluing. Rotate the engine a couple of times; then remove the rocker arm. The bluing should be rubbed off in the center of the stem only. This indicates that the rocker's motion is confined to the center of the valve stem and that it is not running off the edge, where it will push or pull on the valve.

As stated, if your cam has over 0.500in lift, roller-tip rockers must be used. You cannot get the geometry right without them.

Pushrod length may be something you will have to experiment with. It is commonly assumed that you should always go to a longer pushrod, but this is not true in all cases. It depends upon what it takes to arrive at the correct geometry, and you will not know this without trial-and-error checking.

As discussed earlier, if you change rocker-arm ratio you will have to either elongate the pushrod slots in the heads with a Louis tool, or use guide plates.

If you elect to use guide plates, the original slots must be drilled

The rocker arms must be centered in this plane as well. Note that the roller is in the center of the valve stem.

To be able to adjust the guide plates, you may have to enlarge the holes that the rocker studs go through. A die grinder and a carbide bit work well.

out. Use a drill bit as large as possible without cutting into the intake runners. The slots need to be drilled because they will interfere with the new position of the pushrods.

When you install the guide plates, leave the screw-in studs loose enough for shifting the plates around. Experiment until the pushrods are correctly aligned. You may have to enlarge the holes in the guide plates to achieve enough room to shift them properly. Enlarge only the holes that the screw-in studs go through; do not enlarge the slots that guide the pushrods.

When using guide plates with 1.6 rockers, step guide plates must be used to allow for proper pushrod alignment. The stud bosses must be milled down the thickness of the plate, no matter what style you use.

Achieving correct valvetrain geometry takes patience and time, but it is a task that must be done and done seriously if the engine is to be efficient and long lasting.

Engine Assembly

7

By now, you should have all of the shiny, freshly machined parts ready to assemble into a strong-running engine. What you *do* with those parts and how you treat the finished product will determine whether you create a strong, healthy runner, a mediocre engine, or a pile of junk. The choice is yours.

Remember in chapter 1 when we discussed attitude? Now is when a good attitude will pay off. Careful assembly and attention to detail and lots of patience are absolutely required. Cleanliness is also another absolute. We can't all be fortunate enough to have fully stocked workshops, but we can all be careful and keep the parts clean. Just a little dirt can destroy all that precise machine work in a matter of moments. So, go to whatever extremes are needed to have a clean work area. Beg for a spot inside the house if you have to; rent or trade something for a buddy's garage; do whatever it takes.

Block Prep

The engine block *must* be clean. This means that any residue left in the cylinders from the boring and honing operations has to be removed completely. If not, when mixed with the oil in the engine this residue becomes like lapping compound. It doesn't take much imagination to understand what that will do to an engine.

To ensure that your block is clean and ready for assembly, discuss this with your machinist.

Sealer is spread evenly around the core opening in the block.

The core plug is also coated liberally with sealer on the back and sides.

If the shop has a good spray cabinet type of hot tank and it is filled with a good strong solution, the cylinder wall may be sufficiently clean.

When you get the block home, you should wipe the cylinders with WD-40 or any lightweight oil and a white, lint-free towel. Wipe until the towel comes out clean. Hot, soapy water and a large, stiff bristle brush specifically for cylinder cleaning is another excellent way to remove all residue. Be sure to dry the cylinders afterwards to prevent rusting.

This cleaning operation should be the very last thing done before assembly. All thread chasing, anything that will deposit metal or dirt in the block, must be done before this final cleaning.

Core Plugs

If you didn't have the shop install the core plugs and oil galley plugs, this is the time to install them. Coat each core hole with a nonhardening sealer; the kind that comes in a can with an applicator brush works well. Also, coat the back and edges of each plug to help ensure a tight seal against leaks.

We used stock steel core plugs on our engines. Brass plugs are available, but are not recommended for racing use. Brass is a soft material and will not fit the hole as tightly as the harder steel plugs. In everyday driving this should not be a problem, but in a racing situation, the plugs may work loose due to excessive vibrating as well as to increased pressure—often a 22psi radiator cap is used in racing, compared to a 14psi cap for street use. So, for high performance, stick to the stock plugs.

To drive in the plugs, use an old socket of sufficient size. Place the plug, coated with sealer, squarely in the core opening and hammer it in using the socket as a driver. Make absolutely certain that the plug is square in the hole. If not it may leak, and once the engine is in the car it can be difficult to put in a new plug. If there is any doubt as to whether the plug is located properly, remove it and install another one. Never try to reuse a plug that has been installed because it may be distorted.

The cam plug at the rear of the block is installed the same way as the core plugs. Use sealer

A hammer and an old socket are all it takes to drive in core plugs. Make sure they go in straight and are not cocked to one side, or they may leak.

around the plug edges to prevent oil leaks.

At the rear of the block, threaded plugs are used to seal the three oil galleys. Coat the threads of the plugs with the same nonhardening sealer that you used on the core plugs, and tighten the plugs into place. The ends of the oil galleys on the front of the engine utilize drive-in plugs. With a rat-tail file, chamfer the edges of the galleys just enough to remove any burrs or swedges. A 3/8in bolt works well to drive in the plugs. Chamfer the edges so it doesn't cut into the plug. Just as the core plugs, the cam plugs should fit squarely in the holes.

As added insurance against the oil galley plugs coming loose, stake them in place. A masonry chisel, one with four edges, works well. Select one larger than the diameter of the oil galleys, place it against the end of the galley, and give it a sharp rap with a hammer. This produces four perfect swedges around the galley that prevent the plugs from coming loose.

Cam Bearings

If the machine shop did not install the cam bearings, now is the time to do so. It does take a special driver, so if you plan to do a lot of engine work you may want to buy a cam bearing driver. They are expensive, however. You can have a machine shop make one specifically for small-block Chevy rebuilds for about $50.

The cam bearings have different outside diameters and therefore must go into the appropriate bore. The bearing box will tell you which number of bearing goes in which hole. Carefully refer to these instructions to ensure proper bearing placement.

Begin with the rear bearing and work your way forward. Reach all the way through the block from the front with the cam bearing tool to install the rear bearing. This will make it much easier to align the bearing, and will ensure that it fits the bore squarely. If a bearing is cocked, it will bind the camshaft.

Place the bearing with the oil hole in the eleven o'clock position (as you look at the bores with the block turned upside down). This will allow the camshaft to be oiled just before its point of maximum loading. Drive the bearing

After installing the front oil galley plugs, the galleys are staked with a four-point masonry chisel. This is to keep the plugs from working loose.

in so that the oil hole is centered in the middle of the oil groove in the bore, and so that the oil hole in the bearing is fully open. Install the next two bearings in the same manner.

To install the two front bearings, reach through the rear of the block with the tool to help ensure they are installed squarely. Be sure to oil the bores and backs of the bearings to help them slide into place without galling or gouging.

After the bearings are installed, get that new camshaft out of the box and wash it with solvent to eliminate any dust or dirt that may have accumulated during shipping. Dry the cam, oil the bearing journals, apply cam lube, and carefully slide it into place through the front of the engine.

After you have it through the first three bearings you'll notice that it gets hard to handle. Stick a Phillips screwdriver in one of the bolt holes on the front of the cam to act as a handle as you guide it in the rest of the way. Be extremely careful not to damage the bearings with the cam lobes. Take your time, and be gentle. Patience is required here!

If the cam does not fit into the bearings or is tight and difficult to turn over, then remove it. The problem will be either that the cam bores are not perfectly round, or that one or more of the bearings are cocked a bit.

You can remedy this, though. Did you keep your old camshaft? If not, borrow one from a friend or get one from a repair shop. With a file, cut a diagonal groove in each journal. Also, to make it easier to protect the cam bearings from damage, set an old bicycle inner tube over the lobes. You may need to grind down the lobes to get the tube to fit.

Slide this "special" cam into place and rotate it so that the grooves you cut in the journals will peel off the bearing material where it is causing the bind. Remove this special shaft, clean the bearings and block with solvent and air to remove the shavings, and once again try the good cam. If it doesn't turn freely yet, insert the grooved cam and remove some more bearing material. Keep doing this until the good cam slides into place easily and turns freely.

Main Bearings

Once the cam is in place and turns as it should, you are ready to install the crankshaft. The machine shop should have the crankshaft oiled and inside a plastic bag for protection. Remove it and inspect it for any nicks or dings. Shine a light through every oil hole to check for dirt. If you find any dirt or grinding debris, remove it with solvent and a rifle-cleaning brush. Work the brush back and forth through the oil passages.

Now open the main bearings from their packaging and inspect them. Occasionally, bearings will be packaged incorrectly. If your bearings are supposed to be 0.010in undersize, you may find a standard or 0.001in undersize or 0.020in undersize bearing mixed in.

To ensure that all bearings are the same size, measure the thickness of each bearing, both halves, with the dial calipers. This is a quick and easy way to determine conformity. If one bearing measures differently than the others, double-check by measuring again. If indeed the size is incorrect, exchange it for the correct size at the place of purchase. If all bearings are the same, they are ready for installation.

The bearing half with the groove in it goes into the block. The little tang in the bearing fits into the notch in the bearing bore. Do not oil the backside of the bearings. Also, do not remove any material from the bearing surface with Scotch-brite or anything else. You may have heard that the coating on the bearings' surface should be removed. Not true! It is there for a purpose. The bearing manufacturers would have left it off to save money if it was not needed.

The only thing you should do to the bearings is make sure they are free of dirt and dust. Cleaning them in solvent is an acceptable means of doing this.

Install the nongrooved bearing halves into the main caps, again leaving the backsides dry and oil-free.

Bearing Clearance

Next, you need to determine exactly how much clearance there is between the crankshaft journals and the main bearings. There are several ways of doing this.

Plastigage is probably the most common method among backyard mechanics. Plastigage is a small thread-like strip of wax. A small piece of this wax is placed on each crankshaft journal and the main caps are torqued into place. The Plastigage is flattened as the caps are tightened. The main caps are removed and the width of the flattened wax is measured against a scale that comes on the product package. The wider the flattened Plastigage, the less clearance there is. This method can give you a gen-

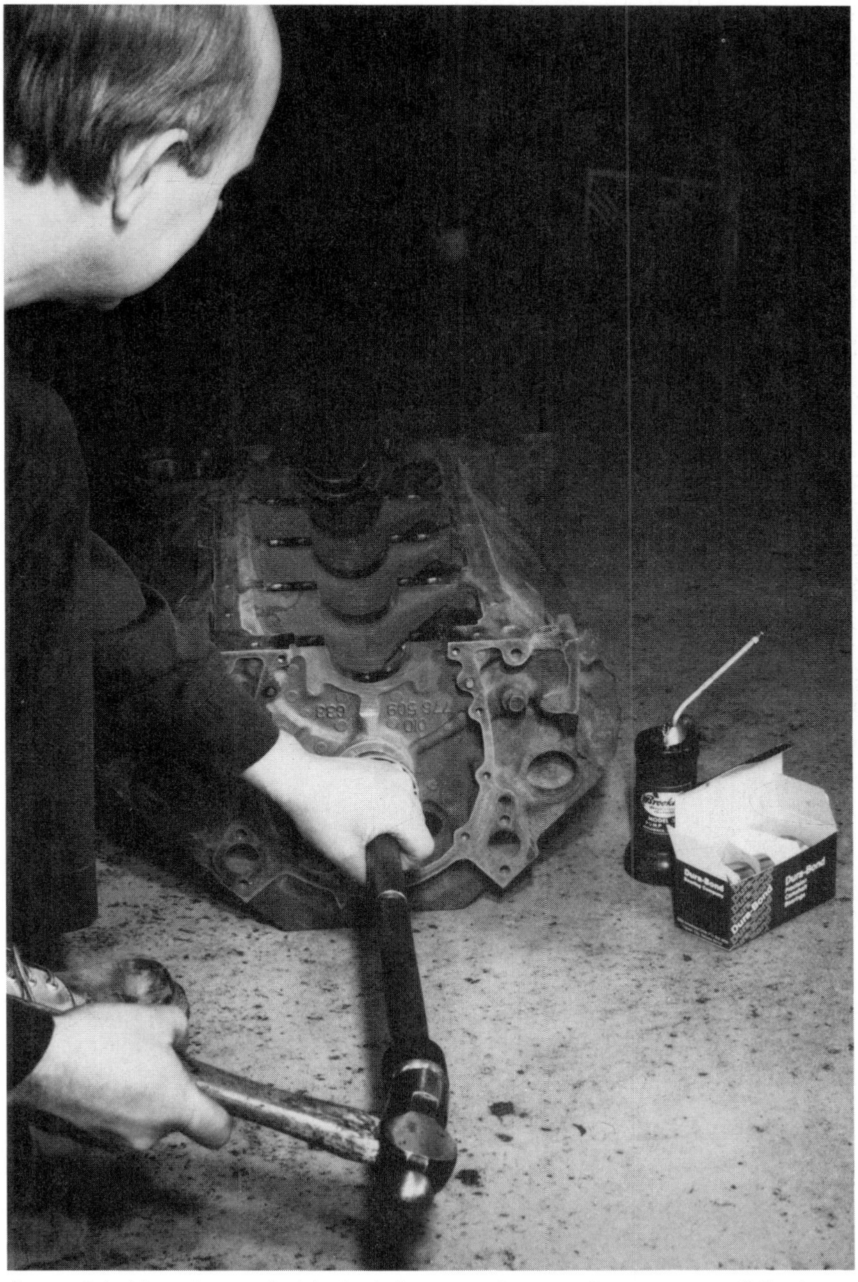

A special driver is needed to install cam bearings. There is no other way to do it! The bearings must be installed in a specific order according to the directions on the bearing box. The inside diameters are the same on all the bearings, though, so it takes only one plug to fit them. Oil the cam bearing bore and the back of the bearing before installing.

The plug has just come through number-two bearing bore. The bearing is well centered now. Though the oil hole in the bearing may be installed in any position and still function, it is best to install it at 11 o'clock as you are looking at the block upside down. This feeds the oil to the camshaft just before its point of maximum loading.

Gently install the new cam after first oiling the bearings and journals. Also, do not forget to apply a liberal dose of camshaft lube to each lobe. This prevents premature wear.

eral idea of clearances, but not exact measurements. Also, Plastigage loses accuracy with age due to its drying out. If you must use it, insist that it is not outdated.

The most accurate method to determine crankshaft-to-bearing clearances is to place all the bearings in their bores and torque the main caps to specification. Do this without the crankshaft in place. Now, with a dial bore gauge measure the diameters of the main bores with the bearings. Compare this reading to the diameter of the crankshaft main journals. The difference in readings is the bearing clearance.

If you don't have a dial bore gauge you can use an inside micrometer. Be sure to measure the inside micrometer with the same outside mic you will use to measure the crankshaft to help guard against error.

In reality most people do not have a set of micrometers, let alone a dial bore gauge. So, to accurately check the bearing clearances in the engine you can arrange with the machine shop to make the check for you. Assemble the bearings in their bores and torque the caps as previously described, and take the block back to the shop along with the crankshaft. They will take the measurements for you. Each shop will approach this differently—they may do it free of charge as an added service to you after doing all the machine work, or you may need to pay for it. Talk it over with your machinist.

Rear Main Seal

Once it is determined that the main bearing clearances are correct, remove the main caps for crank installation. But before the crank is installed, the rear main seal must be put in place.

There are two important factors to remember about rear main seal installation. First is which direction the lip faces. The lip, that part which makes actual contact with the crankshaft, must face the inside of the engine. If it doesn't, an oil leak *will* result.

The second factor is to disregard anything else you've read or have been told about positioning the seal so that the halves of the seal do not coincide with the parting lines of the main cap. This is nonsense; oil will not leak if the parting lines of the two line

A Phillips screwdriver in one of the bolt holes on the end of the cam acts as a handle, making it easier to slide the cam in the last few inches. It is important not to damage the bearings with the cam lobes.

If the cam will not turn easily, remove it and use a special cam like this to true up any bearings that are cocked. This is just an old cam with diagonal grooves cut with a file on the journals. That's an old bicycle inner tube over the lobes to protect the bearings.

Use your dial calipers, or a micrometer if you have one, to make sure all the bearings are the same thickness. Though not common, sometimes a different-size bearing is inadvertently slipped into a set. A dial caliper will spot the problem quickly.

Make sure the main bearing bores are clean and oil free, then install the upper bearing halves—these are the ones with the grooves. Lubricate the bearing surface with oil before installing the crank.

up. If it was necessary to install a seal this way, GM would already be doing it on the assembly line. Put in your rear main seal the correct way, with each half fitting completely in its groove in the cap and the block.

Coat the lip with a thin oil film. This protects it from damage by the crank when the engine is fired up initially. Without this oil film the seal would be burned quickly from the friction and a leak would result.

Crankshaft

Apply a liberal coating of oil to the main bearings, both in the block and in the caps. With the bearings well oiled, gently and carefully lay the crankshaft in place. Once in place, rotate the crank to make sure it is positioned properly and turns freely.

Now install the main caps in the correct order. The rear main needs a little special attention. Apply a light film of sealer to it from the seal on back. This will fill in any irregularities in the mating surfaces to prevent an oil leak. Make sure each cap fits squarely in its place. Tap it down so that one side is not cocked.

When you are sure that each cap is perfectly in place, lightly oil the threads of the main bolts. The main bolts are to be tightened to 70lb-ft, no more, no less. If you don't have a good torque wrench for this job, then buy one, rent one, or borrow one—torque specs, especially on main bolts, rod bolts, and head bolts are not a guessing game.

Go through the bolts and snug them down first. Alternate left and right on each cap so that the cap is drawn down evenly.

When you torque the bolts, do so in two stages, increasing the torque in each stage. For the main bolts at 70lb-ft, set the wrench first for 45lb-ft and tight-

Small-Block Chevrolet Torque Specifications

Main cap bolts (oil on threads)	70lb-ft
Rod bolts (oil on threads)	45lb-ft
Oil pump bolt (oil on threads)	55lb-ft
Head bolts (sealer on threads)	65lb-ft
Intake manifold bolts (sealer on threads except four corner bolts)	35lb-ft
Harmonic balancer (oil on threads)	65lb-ft
Flywheel bolts (oil on threads)	65lb-ft

en each bolt the same amount. Then set the wrench for 70lb-ft and torque each bolt again.

Thrust Bearing

Leave the rear main cap loose. Go ahead and tighten the other four, but the rear cap has to remain free to set the thrust bearing.

To set the thrust bearing, pry the crank forward with a screwdriver wedged between a counterweight and a main cap. With a brass drift and hammer, strike the rear of the crank on the flywheel flange. Tighten the rear cap to spec and remove the screwdriver. The crank should turn freely and have end play, or back-and-forth movement. If not, repeat the procedure, except force the crank toward the rear of the block this time.

Once you have end play, check to see how much. If you have a dial indicator and a magnetic base, this is easily done. Simply position the indicator against the flywheel flange and with a screwdriver pry the crank back and forth, noting the distance on the indicator.

If you do not have an indicator, the check can be made with a feeler gauge. Pry the crank all the way back and check the gap between the thrust face on the crank and the face on the thrust bearing. End play should be 0.004–0.012in.

Doublecheck to make sure you haven't missed torquing any bolts. Spin the crankshaft. It should turn freely and evenly with no tight spots. If it doesn't, back up and look for the problem.

Piston Rings

Piston rings are durable parts—they withstand the pressures and heat of combustion. But, ironically, these durable parts are also fragile; they may be destroyed before they ever make it inside the engine. There are some necessary precautions

The best, most accurate method of determining bearing-to-journal clearance is by torquing the mains down, with bearings installed, then measuring them.

that must be taken when handling and installing the rings.

First, the machine shop will provide the proper cylinder wall finish per the ring manufacturer's suggestions. Second, you should have cleaned the cylinders with hot, soapy water before you began the assembly. These are two important steps to long ring life and proper ring function.

If you do not have a set of ring expanding pliers, buy one. They are inexpensive and are a must for installing rings without doing damage. If you spiral the rings onto the pistons, they will twist and become misaligned. Their axial alignment becomes distorted and creates a "lock washer" effect; the rings will not lie flat. Besides axial distortion, the rings may be radially distorted as well, meaning they will no longer be circular. The result is increased blow-by, which causes poor oil control and performance.

When you use the ring pliers, care must be taken not to overexpand the rings. This will cause them to crimp at the point opposite the gap. It is a good idea to squeeze the ring with one hand while opening the pliers with the other. Then gently lower the ring over the piston and into the groove and carefully release the tension on the pliers.

Stagger the ring end gaps—they should not be lined up. That is really the only rule. If you position the top ring with the gap toward the front of the piston and the second ring 180 degrees around, this is simple to do and works well. Apply the same principle to the three-piece oil ring assembly; place the gap on one rail to one side of the gap in the expander, the gap in the other rail on the other side. The expander gap should be above the pin.

Pistons and Rods

The machine shop will have the pistons installed on the rods as described earlier. Each piston will have an indication of some sort, such as an arrow, a dot, or the word "front" inscribed, to show which direction it goes into the engine. Obviously, the markings face the front of the engine.

Oil the cylinder walls and lightly oil the rings and piston

A dial bore gauge is used to measure the inside diameter of the bearing with the caps torqued. This reading is compared to the main journal size on the crank. The difference is the clearance.

skirt. No need to dip piston-ring assemblies in oil. Use an oil can, squirt a little oil on, and spread it evenly.

To install the pistons in the bores you will need a ring compressor. There are several types on the market. Buy whichever suits your pocketbook and taste. A word of caution here: The rings may actually be damaged by the compressor if you are not gentle with them.

The oil seal must be installed with the lip pointing to the inside of the engine. If not, it will leak!

Once the pistons are installed and you're done using the compressor, the natural tendency is to just toss it onto the bench. Take the time to lay it down gently. When it hits the bench there is a possibility, especially if it strikes another tool, that it will be nicked or burred. Periodically check the compressor for nicks and burrs and, with a piece of emery paper, remove any that exist.

Protect the crankshaft from nicks by installing rod bolt protectors. (Actually, they are crankshaft protectors.) If you don't have any special bolt protectors, then use a piece of hose, such as fuel line hose, that fits snugly over the bolts. No matter how careful you think you can be, you will undoubtably strike the crankshaft with a rod bolt, so don't take a chance—use those bolt covers.

Use the handle of a neoprene hammer to drive the piston down into the bore. Use one hand to reach up underneath to direct the rod over the corresponding journal. Once in position, install the appropriate cap. The bearing tangs both go on the same side; also, the machine shop will have numbered the rods and caps—the numbers are on the same side of both rod and cap.

When all eight piston-and-rod assemblies are installed, torque the rod bolts, with oil, to 45lb-ft. There has to be some side clearance between each pair of rods, but not too much. Excessive clearance will cause a noise in the engine. A feeler gauge is used to check side clearance, which should be 0.004–0.020in. Whatever gauge will fit snugly between the pair of rods is the clearance. If there is no clearance, the sides of the rods can be ground to provide it. If there is

Do not do this! See how the seal halves are staggered? This does nothing to prevent oil leaks. Install the seal as it was meant to be installed, and forget about any of these "tricks."

excessive clearance, then you'll have to try another pair of rods that are thicker. This is not a common problem, but at the same time, clearance has to be checked.

Rotate the crank-and-piston assembly through a couple of complete revolutions to make sure it turns freely and smoothly. If it binds or gets tight in one spot, find out why.

Oil Pump

If you are using a stock-type pump as we did on the 355, it will come without a pickup. You may buy a new one, but, when you tore down the engine the old pickup should have gone to the machine shop for cleaning.

Before you install it, check carefully for dirt and chunks of oil seals and nylon timing gear teeth. Use a flashlight to inspect the screen, paying careful attention to the area that is hard to see underneath the shield. Rap the pickup on the bench to help dislodge any pieces of trash. Blow through the tube with compressed air. You can't be too sure here—any little particle of dirt or debris could come loose and find its way through the engine.

To install the pickup on the new pump you need either a special installation tool or a 5/8in wrench and a hammer. Start installing it by hand, then with either the tool or the wrench positioned against the flange on the tube, drive the pickup into the pump.

Now, you need to know where the pickup should be in relation to the bottom of the oil pan. To determine this, take a ruler and measure the depth of the pan. Set the oil pump in place, without torquing it down, and position the pickup so that it comes within about a quarter inch of the floor of the pan. You can eye-

Special Information for the 406

There is one vital check to make on the 406ci engine when using the 350 rods, as we did here: Check the clearance between the rod bolts and the camshaft and the rod bolts and the block. *You* have to check this. Failure to do so will lead to destruction of the engine! Rods 1, 5, 2, and 6 are critical to check for clearance with the cam.

Make this check before the pistons are installed. The easiest way is with an inspection mirror. Place a piston in its appropriate bore. Then use the mirror in the opposite bore to look at the bolts and determine clearance.

In addition, all bolts should be checked for clearance with the pan rail. If there is insufficient clearance between bolt and cam and/or bolt and rail, grind the bolt. Grind and check until you get 0.050in clearance. A time-consuming job, but a must.

Apply a bit of sealer to the rear main cap in the area alongside the seal. This will fill in any irregularities in the metal and prevent leaks.

Tap the caps into place. Make sure the threads on the bolts are clean and lightly oiled. Don't forget to oil the bearing half in the cap, too.

Use a torque wrench and torque the bolts to 70lb-ft. This is essential. Borrow a wrench or rent one, but use it and don't go beyond the recommended specs.

To set the thrust bearing, lever the crank forward with a large screwdriver wedged between a counterweight and a main cap. Drive the crank forward with a brass punch and hammer.

ball from the ruler, which you place on the pan rail of the block, to the pickup since the distance does not have to be precise within thousandths of an inch. Move the pickup until it is in the proper position and remove the pump from the block.

Do not braze the pickup to the pump! You may have heard about doing this or even read about it, but, again, do not do it. The heat can distort both the relief spring inside the pump and the pump itself. Also, it is nearly impossible to braze to cast iron, and this is what the pump is made of.

The simplest and safest way to secure the pickup is to use a set screw. Before the pickup is installed, drill a hole in the pump where it will intersect with the pickup tube, then tap it for a set screw. We used a 1/4in set screw, so the appropriate drill size was 13/64in. Rinse out the pump with solvent to remove any metal from drilling and tapping, then install the pickup.

After it is positioned for pan clearance, put in the set screw. Use a drop of Loc-tite on the

After setting the thrust bearing, torque the rear main cap bolts, and rotate the crank—it should turn freely. If it doesn't, repeat the procedure. You may have to drive the crankshaft toward the rear of the block instead of the front. It is trial-and-error. There should be about 0.004–0.012in end play in the crank. You can check this with a feeler gauge as shown here. Lever the crank back with a screwdriver and see how much clearance there is between the crank thrust face and the thrust bearing.

screw for insurance and tighten it down against the pickup tube. This is an easy, safe, inexpensive, sure-fire method of securing the pickup.

Before you torque the pump to the rear main cap, you must determine if you have the appropriate length bolt—there is more than one length for oil pumps and there are two styles of rear main caps. On one style of rear main cap the bolt hole for the oil pump is open: it goes all the way through the cap. It is possible, then, if you use an oil pump bolt that is too long, to screw the bolt into the rear main bearing. If the bolt is too short, then there will not be enough threads engaged to safely hold the pump in place. Thus, the object is to make sure that the bolt is not too long, yet is long enough to engage the proper amount of threads.

To determine if the bolt is the right length, install it in the rear main, *finger-tight only*, without the pump, until all the bolt's threads are engaged. With the dial calipers, measure the length of the bolt that remains above the cap. Compare this with the thickness of the pump. If the length of the bolt is greater, then the bolt is too long. The threads will be engaged, but the pump won't be tight. If the length of the bolt is significantly less than the thickness of the pump, then an insufficient amount of threads will be engaged when the pump is installed. The distance of the protruding bolt should be only slightly less than the thickness of the pump for a good, securely fitting oil pump.

Secure the oil pump sleeve and distributor driveshaft. Use the stock plastic sleeve—there is no reason to spend the money for a steel sleeve. With today's lightweight oil, it simply is not needed.

Once you have found a bolt of the proper length, oil the threads and tighten down to 55lb-ft.

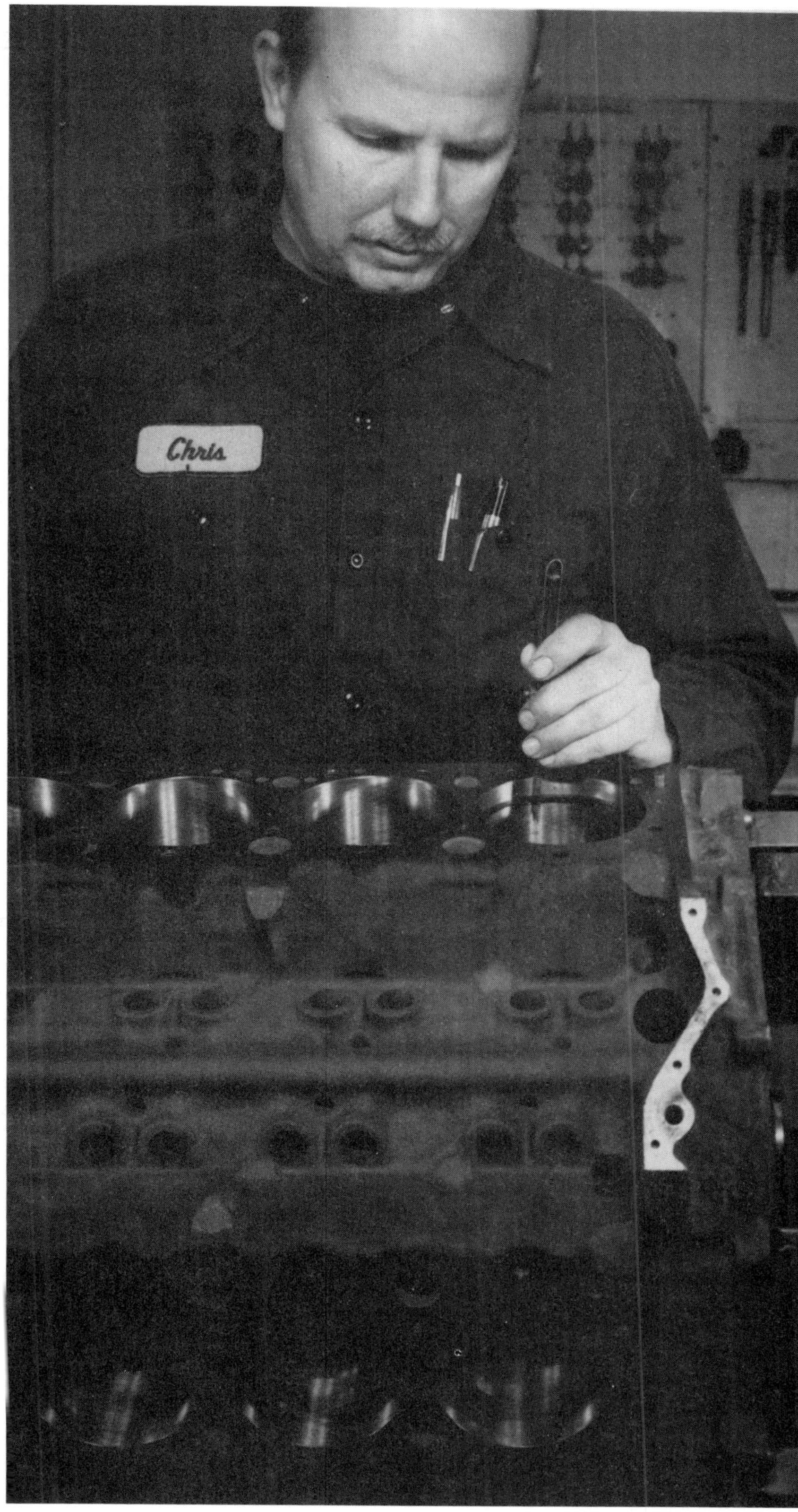

Check each and every ring to make sure they are the right size. Sometimes an odd-size ring gets packaged in a set. Slide the ring squarely into the bore and measure the gap with a feeler gauge. There should be about 0.016–0.024in on the top ring, and 0.012–0.020in on the second ring.

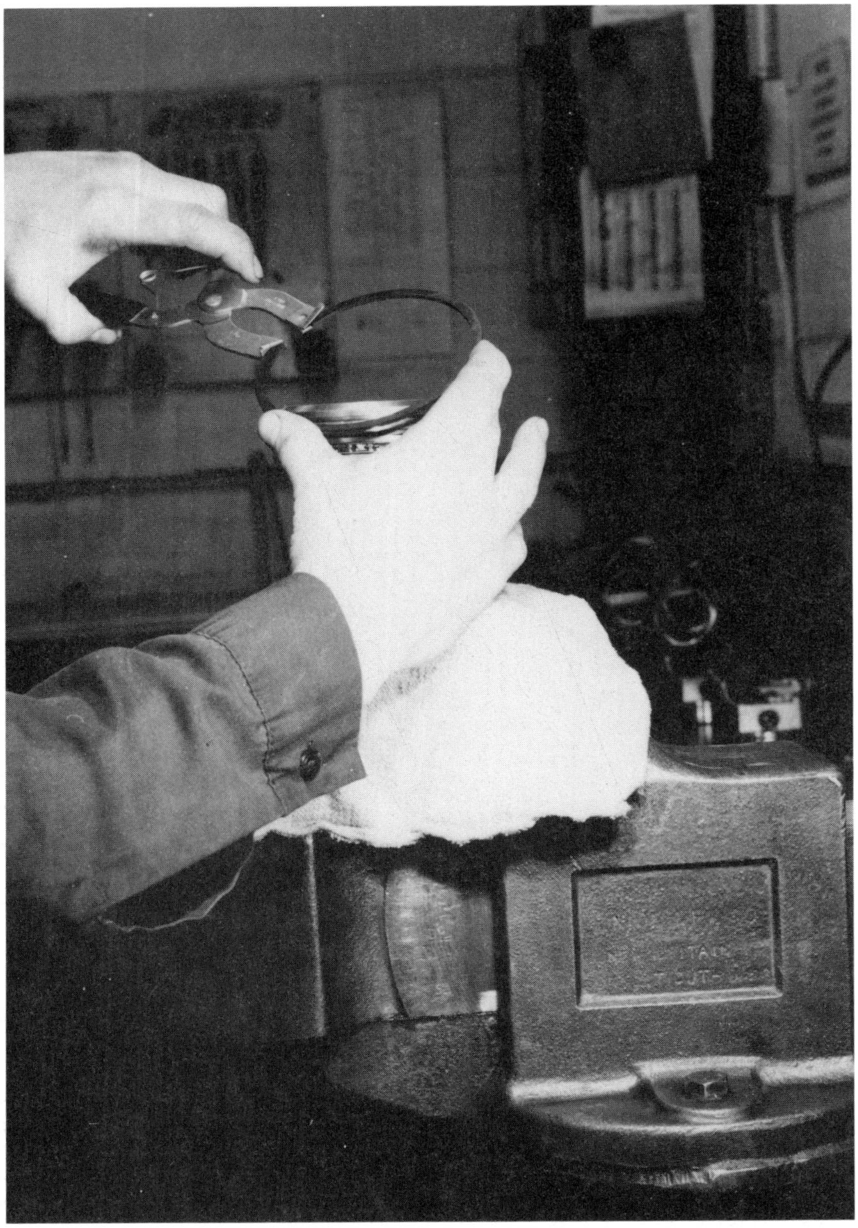

The *only* way to install rings on pistons. Using ring pliers, gently open the rings, being careful not to overexpand.

The hand placed opposite the pliers helps prevent overexpanding.

Timing Chain and Gears

Install the crankshaft gear using a hammer and a piece of exhaust pipe tubing large enough in diameter to fit over the end of the crank. The gear will be snug, but should go on smoothly and without excessive force. Be sure to oil the crank and the gear before driving the gear in place. *Any* part that has a press fit has to be oiled to prevent galling. Now, the camshaft gear and timing chain will go on together.

When installed properly, the dots on the crank gear and the cam gear will be lined up exactly. By this we mean with the engine right side up, the dot on the crank gear will be at the top and the dot on the cam gear will be at the bottom. If the dots do not line up perfectly, shift the cam gear until they do. When everything is positioned correctly, bolt the cam gear in place with the three bolts.

If you are using a timing set that has a multi-indexed crankshaft gear, that is, if there are three positions in which to install the crank gear, install it straight up. One position retards the camshaft timing four degrees, one position advances it four degrees, and the third position (straight up) places it in a neutral position.

It is best to install the cam straight up and see how the engine runs, then decide if it needs advancing or retarding. Advancing the cam timing normally produces more low-end power and retarding it normally produces more top-end power. (See the preceding chapter on cam degreeing and valvetrain geometry for more details.)

Tighten the three bolts that hold the cam gear to the camshaft and double-check the timing mark alignment. If you are not going to degree the cam, you are ready for the timing chain cover.

Use some nonhardening sealer in the front seal bore in the timing cover to help prevent leaks. Position the seal in place. When you drive it in, you need to spread out the force of the hammer strikes equally around the seal so that it will go into place properly. A piece of metal large enough to cover the whole seal works well. Tap the center of the metal plate with a hammer and the seal will be driven in equally all the way around. If you don't do it this way, the seal may become cocked in the bore and cause leaks or damage to the seal.

Set the front cover gasket in place over the two locating dowels using sealer on both sides. Lightly coat the lip of the seal with oil and slide the cover over the end of the crankshaft. Line up the holes and tighten it in place.

A word of caution: Beware of some aftermarket covers; they do not fit well and some even have the seal bore off location so far it doesn't line up with the crank.

Oil Pan

With the short-block assembled, the pan may be installed. Take one last look and recheck the main and rod bolts and the oil pump bolt. Make sure there is nothing lying on top of the engine, such as shop rags and wrenches.

After everything checks out, the pan can be installed. Before bolting it on, make sure it is flat. Lay it on a flat surface and see how much it rocks. A little is all right, but anything excessive and you will have to find another pan (or return the pan if it is new).

If you are installing a used pan, the dimples will have to be taken out of the rail. These are caused by the bolts pulling on the pan when they are tightened. Lay one rail of the pan on the bench, with the sump hanging over the edge. With the hammer, tap the rail to flatten the dimples. Don't squash it, but hit it hard enough to flatten it. Do the same to the other rail.

Aftermarket pans tend to be out-of-flat and the holes do not line up well with the bolt holes in the block. Before putting the gaskets in place, try the pan to see if the holes line up. If they don't, there are two options. First, start one bolt and with a punch or screwdriver in another hole, lever the pan until you get another hole to line up. Continue this way until you get all the bolts started. If some fail to fall in alignment, you may have to drill the holes larger (in the pan).

The pan we used on the 406 is a definite exception to the rule. The large sump, circle-track Champ Pan made by JR Manufacturing, Inc., is a first-class piece of work. It is constructed of heavy-gauge metal, thus providing strength and rigidity. The

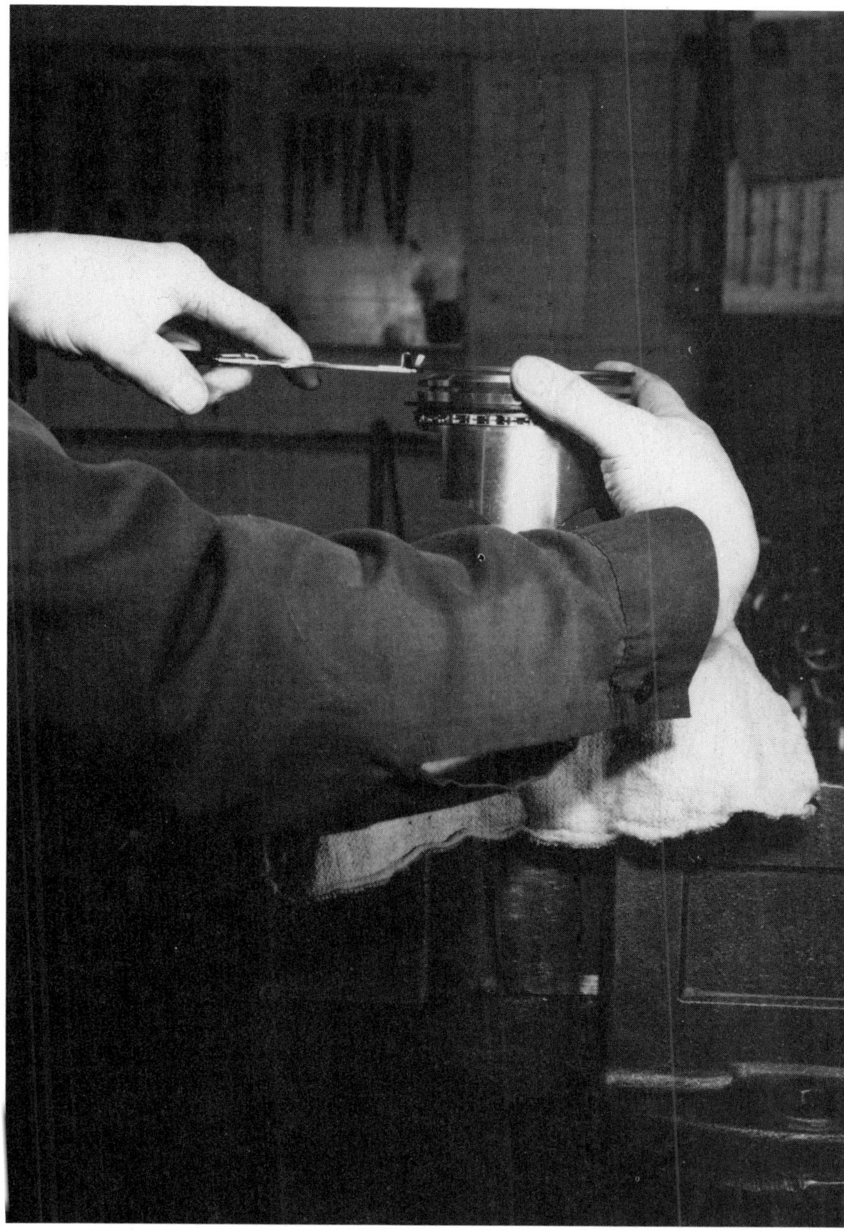

Lower the ring over the piston and into the correct groove. Slowly release the ring pliers.

rails are flat and the holes line up perfectly with those in the block. The pan comes with a special pickup that is pre-set for the pan. All this for a reasonable price, too.

Once the pan is ready for installation, the gaskets first must be positioned on the block. The gasket set may contain more than one set of pan gaskets; determine which ones are appropriate for your application.

Use sealer, such as 3M Weather Stripping, on the block *only*. The purpose of the sealer is not to prevent leaks, but to hold the gasket in place while the pan is being installed. Apply a light coat of sealer, lay the gaskets on, and make sure they are positioned correctly with all holes lining up.

Stagger end gaps like this. The two upper rings are 180 degrees apart. All that really matters is that the gaps are not in line; rings rotate on the piston, anyway. Stagger the gaps on the three-piece oil ring in similar fashion.

Let the sealer dry completely before installing the pan.

Next, set the rubber end seals in place: one in the groove in the rear main cap and the other in the front cover groove. The gasket kit will have two front seals; check the seal length according to the instructions in the kit to determine which one to use. Make certain that the tab in the pan gasket fits into the cutout in the rubber seal. (The pan gasket goes underneath the rubber seal.) Apply sealer at this intersection where the pan gasket and seal meet. A bead is all that is needed—don't get too carried away.

One last glance to make sure there is nothing left behind inside the engine, and the pan is lowered carefully into place. Start all the bolts and tighten

Oil the cylinder walls and the rings before piston installation. Be sure to place the protective boots on the rod bolts—you don't want to nick that fresh crank. Use the ring compressor and slide the piston into the bore. Tap it into place with a soft hammer handle and guide the rod over the journal.

Place the matching cap over the rod bolts and torque to spec.

The easiest way to check for rod bolt-to-camshaft clearance—this is an absolute must for the 406ci engine using 350 rods. There has to be a minimum of 0.050in between bolt and camshaft. Notice that the cap is not installed; you may have to remove the piston and grind on the rod bolts several times to achieve clearance. Be sure to check for clearance between the rod bolts and the sides of the block, too.

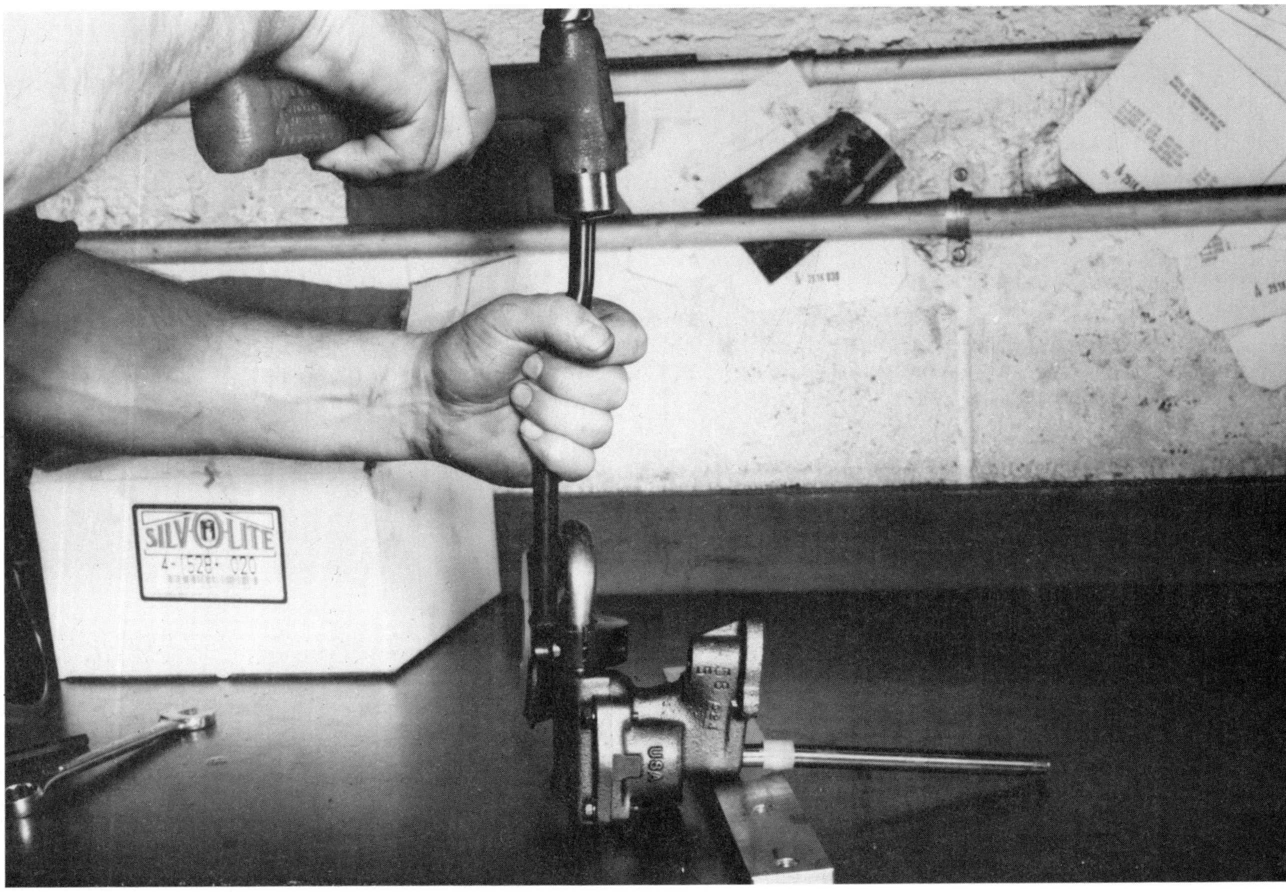

The special oil pump pickup installation tool.

Another oil pump pickup installation tool. This one is simply a 5/8in open-end wrench out of your toolbox. It is cheap and does the job.

them gradually, working your way around. Begin at the center and work toward the ends, alternating back and forth from one side to another. Snug each bolt slightly. Once you have made your way around through all the bolts, go through them again, snugging them a little more this time.

During this process you should go around a total of four times, tightening the bolts a little more each time. This is because the gasket compresses as the bolts are tightened. If you see the gasket start to squish out between the pan and block, then you are torquing the bolts too tight.

Harmonic Balancer

Harmonic balancer installation tools are made, but instead of spending the money, go to the hardware store and buy a bolt about 1 1/2in longer than the stock balancer bolt. Be sure the thread size matches.

Oil the crank and the balancer inside the bore and out. Place the balancer on the end of the crank, oil the threads on the bolt, and start it in. Be sure you have the large washer on the bolt. Tighten the bolt down, pulling the balancer on. Go only as far as it takes to be able to start the stock bolt. *Do not bottom the bolt in the hole*. This will damage the threads on the end of the bolt and in turn will damage the threads in the crank when you remove the bolt. Be sure to oil the threads on the stock bolt. Tighten it down to 65lb-ft.

Cylinder Heads

The short-block is now ready for the cylinder head assemblies. Rotate the engine on the stand so that one deck surface is facing up. Lay a head gasket on the deck of the block, positioning it over the two dowels at each end. The gasket will snap into place over the dowels, which prevents it from slipping.

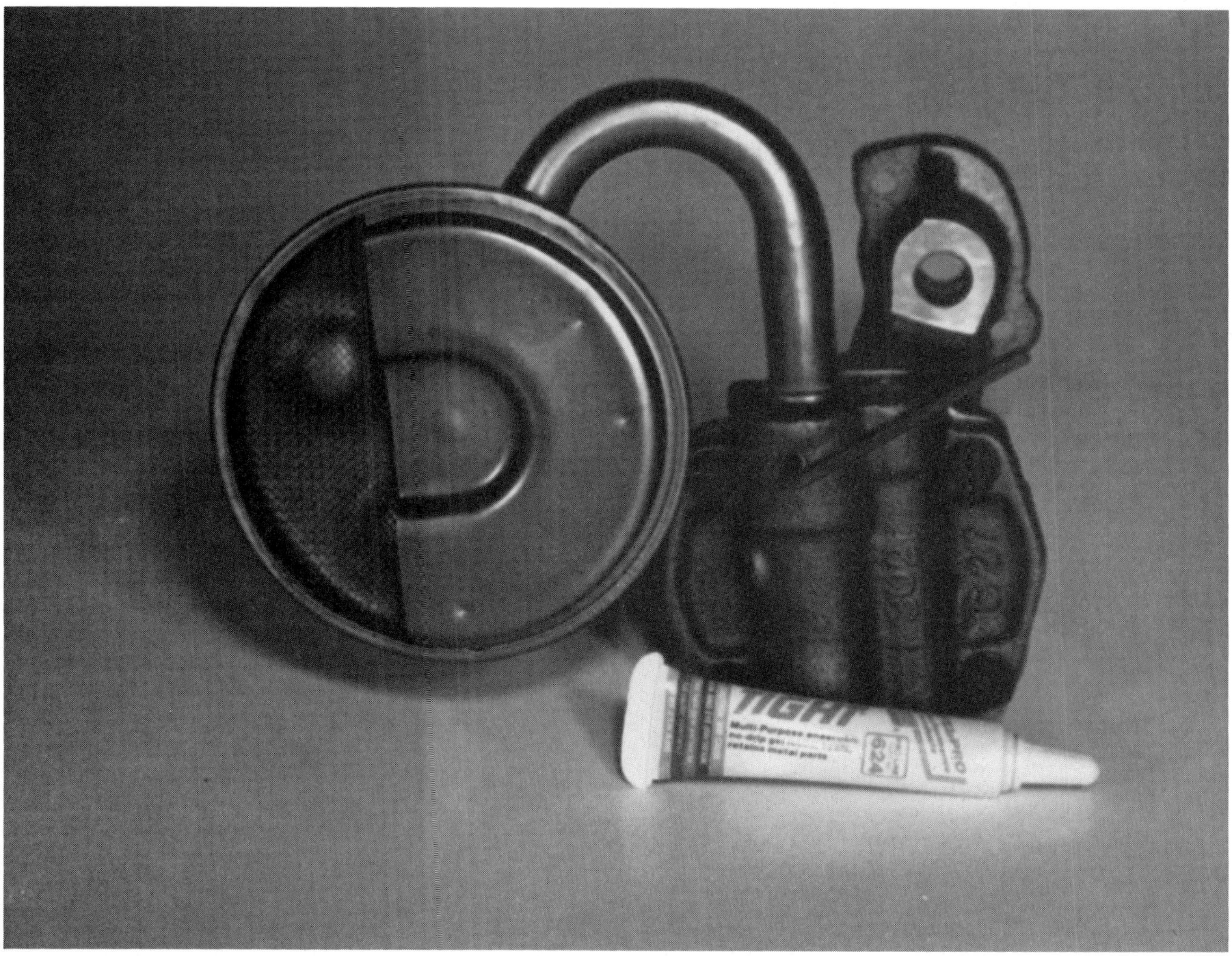

The hardware used to permanently attach the pickup to the pump. Drill and tap a 1/4in hole in the pump, apply Loc-tite to the set screw, and screw it in. Easy, safe, cheap!

It is critical to use the correct oil pump bolt. One method for checking is to install the bolt, finger-tight, and measure the amount of space left between the shoulder and the pump flange on the main cap.

Make sure you have seventeen head bolts laid out, ready to go. There should be seven long bolts, eight short bolts, and two intermediate bolts per head. Since the head bolts enter into the block's water jacket, sealer must be used on them. Copper Coat is excellent for this. Brush a little on each bolt, just enough to coat the threads. Place the head on the block, positioning it over the dowels, and start a couple of bolts to secure it.

When you have all seventeen bolts in place, refer to the torque sequence to tighten the bolts. If you are in a situation where you do not have a diagram to follow, just remember to begin at the center and work out toward the ends in a circular pattern.

The final torque on the head bolts is 65lb-ft. Work up to this spec by going through the sequence at least three times, each time increasing the torque by 15lb-ft. Start with 35lb-ft, go to 50, and finish at 65. You should go through the sequence one final time to ensure that none of the bolts have loosened. Often, the inner bolts will loosen as the outer ones are tightened. Do not be satisfied until each bolt is at 65lb-ft.

If you are using a composition gasket or a permanent torque-type gasket, do not apply sealer. If you use a steel shim gasket, as we did on the 355, apply Copper Coat to both sides and around all bolt holes.

Lifters, Pushrods, and Rocker Arms

Clean the lifters with solvent and blow them dry. With an oil can, squirt oil into each one through the bleed hole on the side. This applies to both solid and hydraulic lifters. Don't worry if they aren't filled with oil; when the engine is primed, they will fill. Now, using the cam lube that is left over from installing the camshaft, liberally coat each lifter and place it in a lifter bore.

Compare this measurement with the thickness of the pump where the bolt goes through. The bolt should be slightly shorter to ensure that the pump will be tight and the maximum amount of threads will be engaged in the cap.

Instead of measuring, you can get a close idea by holding the pump next to the bolt as it is installed.

Even if you are using new pushrods, roll each one across a flat surface to make sure it is not bent. If they all check out, slide them down through the pushrod slots in the heads and center each one in its appropriate lifter.

Oil the tips of the valve stems and the rocker studs, then place the rockers on the studs. On stock-type rockers, oil the rocker balls and put them in place, then start on each nut. A drop of oil should be applied to the top of the stud after the nut is started. Simply make sure everything in the valvetrain is well lubricated since there is a tremendous amount of friction on these parts during the initial fire-up.

Make sure the pushrod is centered in its lifter and also in its place in the rocker-arm socket before you tighten the rocket arm.

Setting Valve Lash

On hydraulic lifters, as in the 355ci engine, there is zero lash. The plunger in the lifter has to be preloaded, though.

Setting lifters, whether hydraulic or solid, is actually easy. Forget any of the complicated systems you have read about elsewhere. Simply work with one cylinder at a time and set both valves. Begin with number one cylinder because it is at the front of the block. Rotate the engine until the exhaust valve just begins to open. Stop. Now set the intake valve. Rotate the engine again until the intake valve, on the same cylinder, is about to shut. Stop. Set the exhaust valve. Now, move on to the next cylinder down the line and repeat the procedure.

It is that simple. Write down this procedure on a piece of paper and tape it inside your toolbox or carry it in your wallet so you will have it handy.

To adjust hydraulic lifters, with the engine in the correct mode for the particular lifter you are adjusting, tighten the rocker-

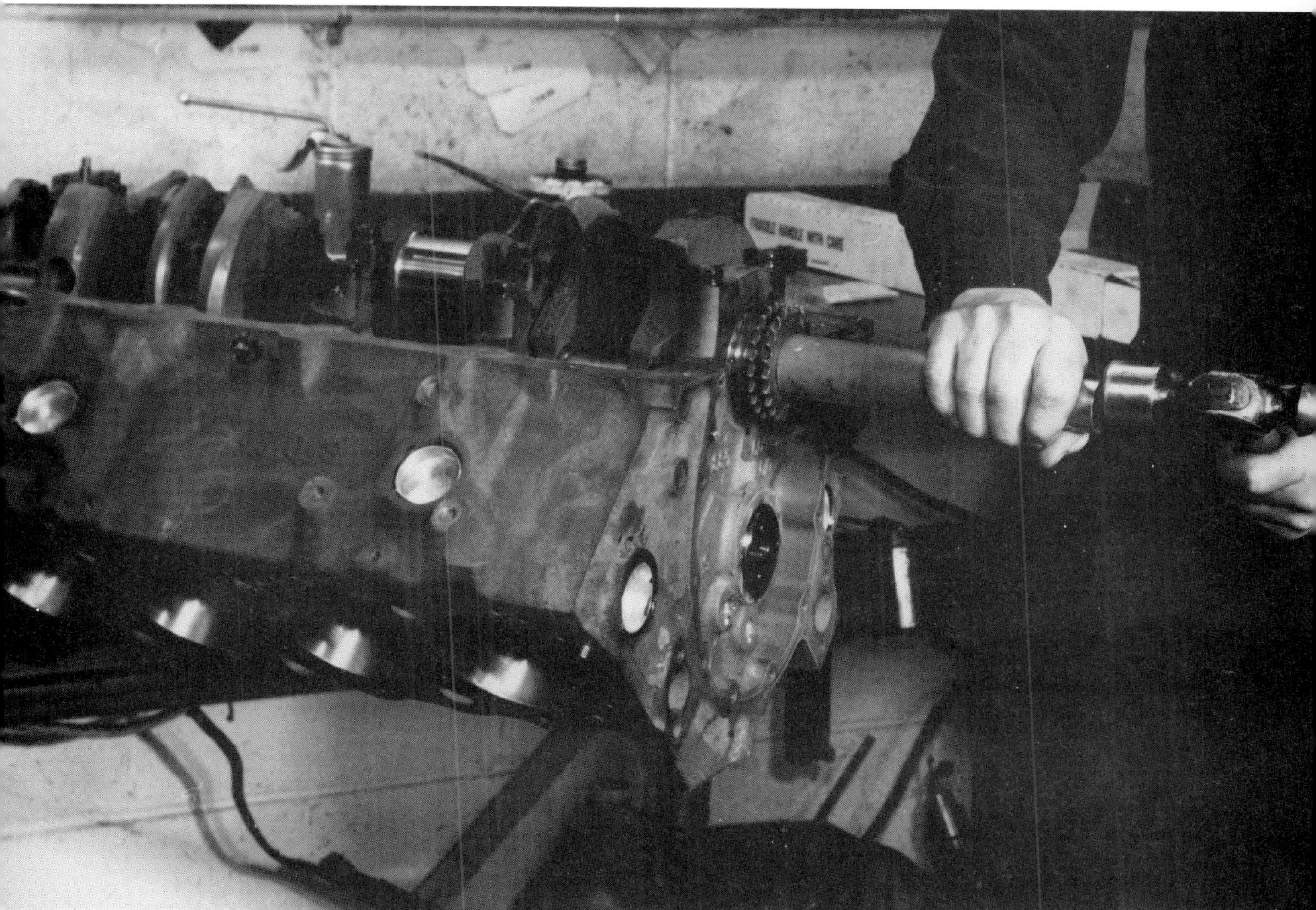

You don't need a special tool to drive on the crank gear; just use a piece of exhaust pipe that fits over the end of the crank. The gear will drive on easily since the press fit is minimal.

arm nut while spinning the pushrod between your fingers. When the pushrod stops turning, that is, when all the lash is taken out, turn the adjusting nut another one-half to three-quarter turn. The lifter preload is set. Move on to the next one and repeat.

On solid lifters, refer to the cam spec card for the valve lash—this is the distance between the tip of the rocker arm and the tip of the valve stem. Follow the procedure just described, but instead of taking out all of the lash, use a feeler gauge between the rocker and valve stem to arrive at the specified lash. If the spec calls for 0.024in lash, a 0.024in feeler gauge should fit snugly between the rocker and valve stem. There should be drag on the gauge as you draw it through. Take your time and double-check to make sure the lash is set accurately.

If you use 1.6:1 rockers, add an additional 0.002in to the recommended clearance. If you do not do this, the higher-ratio rockers will increase the cam's duration, which in turn raises the effective power range of the engine.

Intake Manifold

Before installing the intake manifold, check inside the lifter valley for left-behind tools and shop rags. It is easy to lay something down and forget it, so double-check.

Use either 3M Weather Stripping or contact cement to glue the intake gasket to the cylinder heads. Apply a thin coat of the cement to the gasket, on the side that will go against the head, going around the water passages and intake ports. Lay the gaskets on the heads, make sure they are positioned correctly, and press

Line up the timing marks and install the chain and cam gear. The dots on the two gears should line up exactly, as shown. If you hold a straightedge through the centerlines of the crankshaft and camshaft, the dots should fall in this line.

A bolt in the end of the crank with a nut on it makes a cheap, convenient tool for rotating the crank during assembly.

Apply sealer to the front oil-seal bore in the cover.

them down firmly. Let the cement dry for at least 30 minutes before proceeding any further.

The primary function of the cement is to hold the gaskets in place when you put on the manifold. If the gaskets are free to move around, they no doubt will cause an intake leak and/or a water leak.

Your gasket kit will probably come with rubber end gaskets for the intake manifold. Disregard these; instead, use RTV sealer. The surface must be free of oil for the sealer to stick, so clean the area with solvent. After waiting at least 30 minutes for the cement on the gaskets to dry, put a dab of RTV in each of the four corners where the heads and block meet. Work the sealer into the corner with a fingertip making sure it is pushed all the way down into the crevice. Run a thin bead around the water passages, then run a bead about 1/4in thick along the rail from one head to the other. Do this on both the front and rear. Carefully lower the manifold into place.

The manifold bolts should be torqued to 35lb-ft. It is impossible on some manifolds to get a torque wrench on all of the bolts. All you can do is make a sensible guess at how tight you are getting the bolts. With a box end wrench or a short ratchet you will not overtighten—just do not use a "cheater" pipe on them or you will end up with problems.

If you cannot use the torque wrench, but are concerned about being as accurate as possible on the 35lb-ft, torque a 3/8–16in bolt (this is the same size as a manifold bolt) into something else, such as a cylinder head, with the torque wrench. Pull on the bolt with the wrench you will use on the manifold to get a feel for 35lb-ft.

Work your way from the center of the manifold outward as you tighten the bolts. Go around them several times to make sure they are all tightened evenly. As with the other bolts on the engine, tighten the manifold bolts in steps, going tighter each time.

There is one important thing to watch for here. On each side of the manifold there is one bolt that can hit the pushrod if it is too long. Look into the bolt holes with a flashlight. One of the two center bolt holes will just about line up with a pushrod. In this hole, a shorter bolt must be used so that it won't reach the rod.

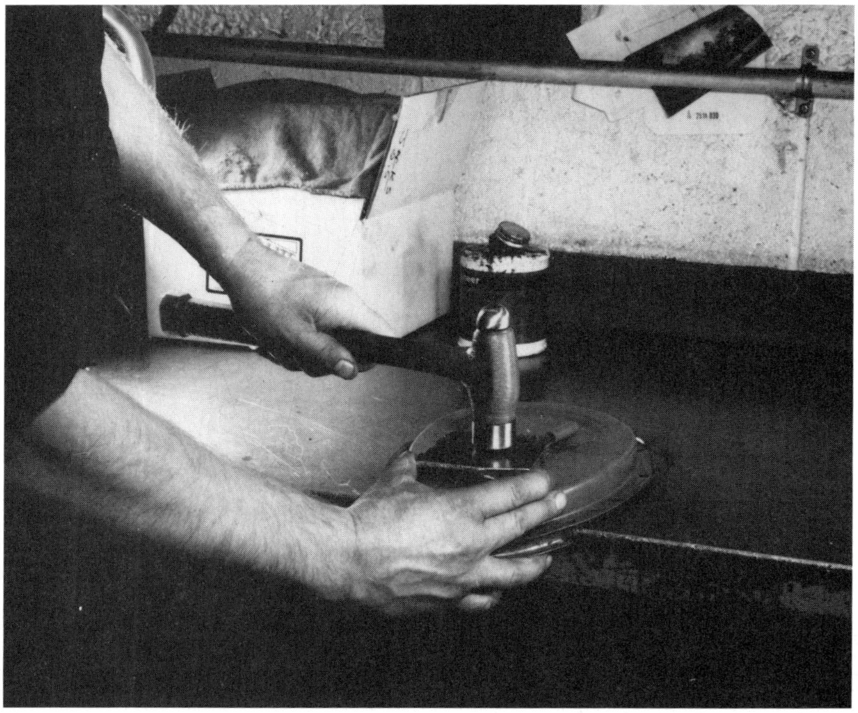

Apply sealer to the outside of the seal and drive it into place. You need to evenly distribute the driving force—a plate of steel large enough to cover the entire seal works well.

Valve Covers

Flatten out the dimples around the valve cover bolt holes as you did on the oil pan. Clean the gasket area on the covers with solvent or Brake Kleen.

After the covers have dried, glue the gaskets in place with 3M Weather Stripping or contact cement. Set the covers on the heads and snug them down. Don't get carried away here—they just need to be snugged.

Finished Product

Now the engine assembly is completed. You can step back and look at your handiwork on the engine stand. It is a good feeling to be able to observe your accomplishments, especially when they are tangible. You are probably anxious to hear it run—this is the *real* test of your work. A smooth-running, powerful engine can give you reason to be proud.

Taking the dimples out of the oil pan. A pan that has been used will dimple around the bolt holes. Lay it on the edge of the workbench and tap the length of the rail to smooth it out.

Use 3M Weather Stripping to hold the pan gasket in place. This is the only purpose for the weather stripping; it does not act as a seal. Let it dry completely before installing the pan.

Install the pan gaskets and the rubber end seals that go in the rear main cap and the front cover. Use sealer at the intersections where the cork gasket meets the rubber seals. Place the pan in position and tighten it down.

If your harmonic balancer is damaged on the seal surface, as this one is (see the groove worn in it), it can be repaired with this speedy sleeve.

Lube the lifters by pumping oil in through the bleed holes on the side. Don't be too concerned if you can't get them to fill up. When you prime the engine, they will fill.

Always use cam lube on the lifters. Apply a liberal amount before installing them.

At this point, the engine will look like this: the heads are on and torqued, the lifters are installed, and the pushrods are in place.

As you assemble the valvetrain, oil each piece well. There will be a lot of friction on these parts when the engine is first fired up. Oil the threads on the studs, and oil the tips of the valve stems and pushrods.

Add oil to the tops of the studs after the nuts are in place. Be sure to lube the rockers and the rocker balls with cam lube—there is a lot of pressure here and it will wear new parts in a hurry.

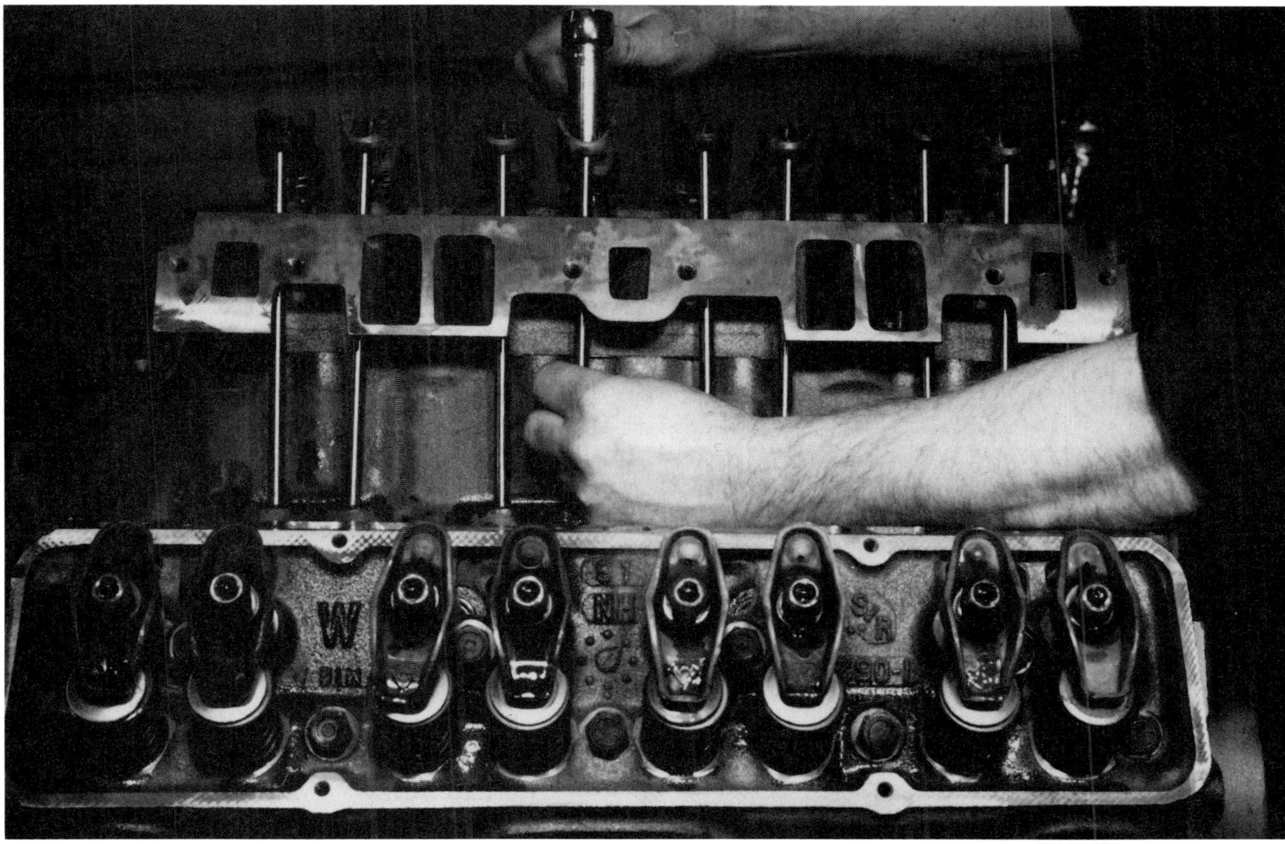

To adjust a hydraulic lifter, tighten down the rocker nut until the pushrod stops turning; then go down another half to three-quarter turn. This sets the preload on the lifter. Make sure the pushrods are centered in the lifter and positioned properly in the rocker arm.

Instead of rubber gaskets, use silicone to seal each end of the manifold. Notice that the intake to head gaskets are in place already. Lower the manifold into place and tighten it down.

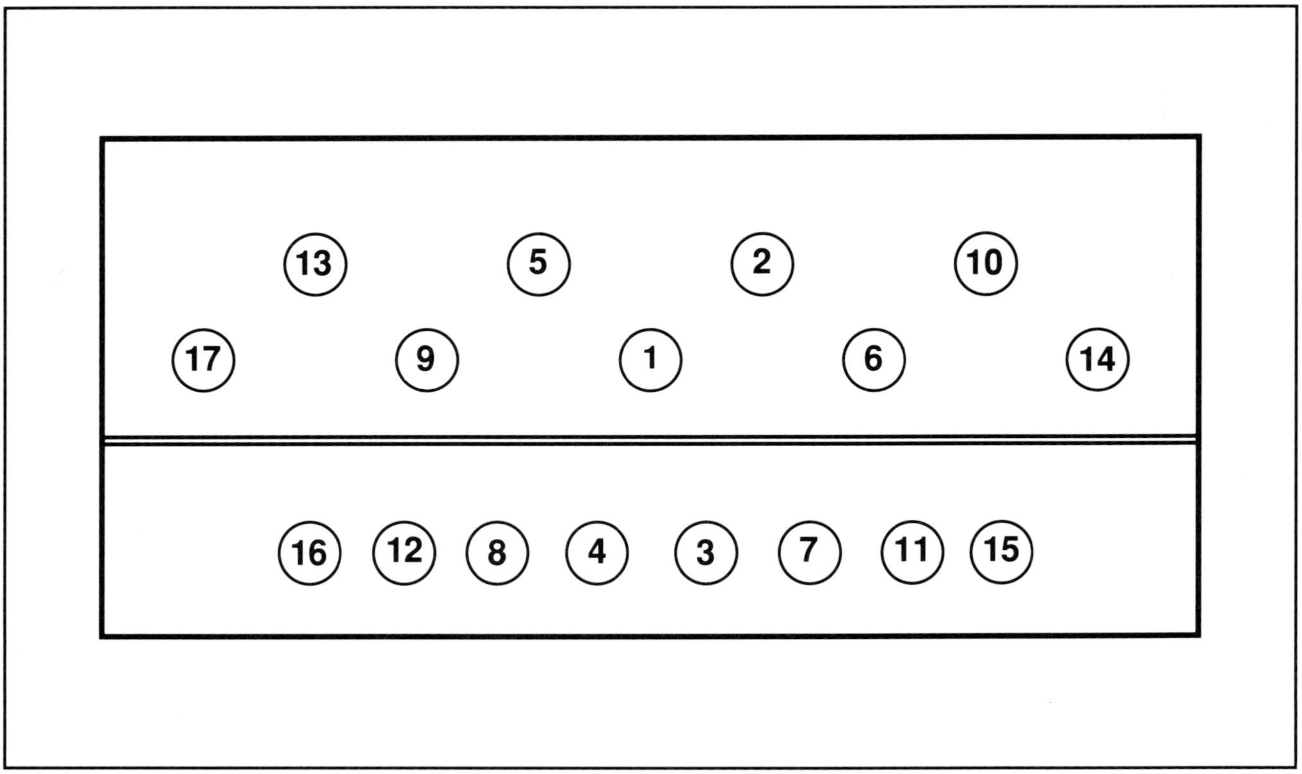

Diagram 1
Head bolt torque sequence.

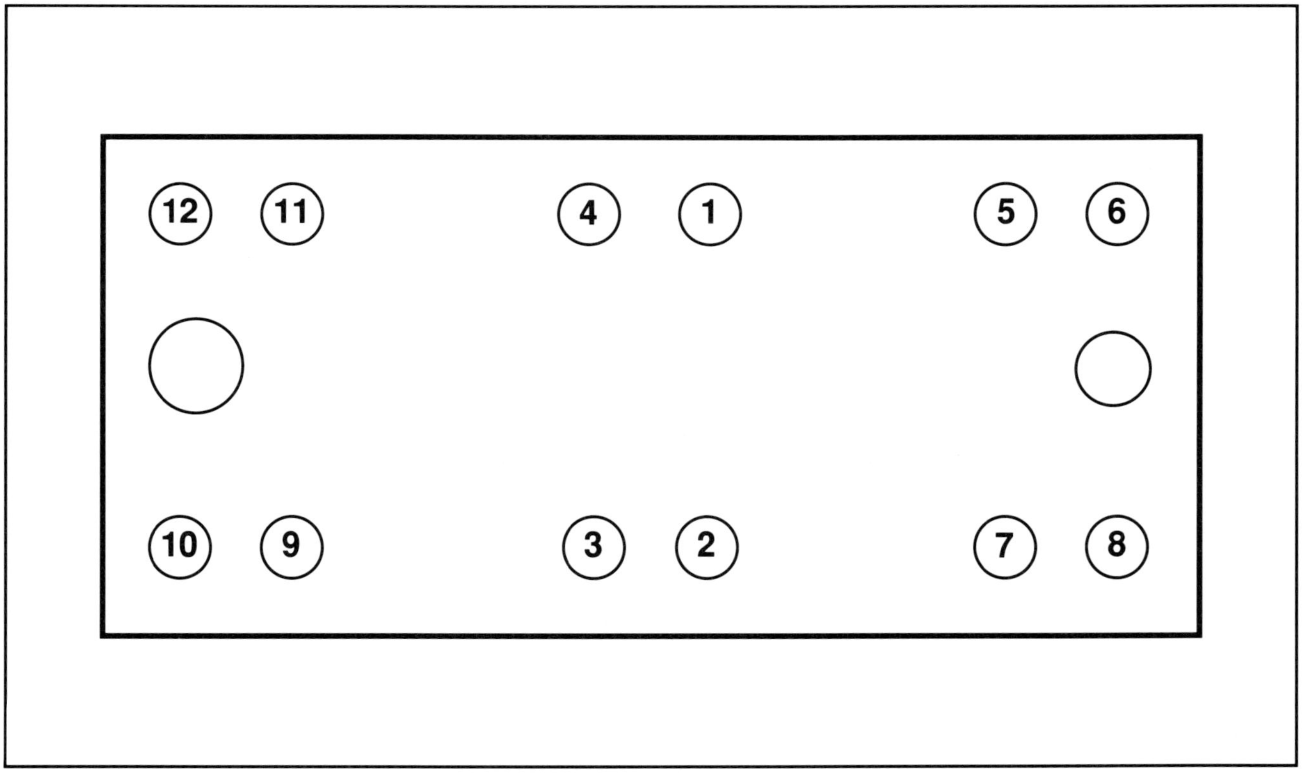

Diagram 2
Manifold bolt torque sequence.

Engine Fire-Up

8

Priming the Oil System

Before the engine is started for the first time, the oiling system must be primed. Fill the pan with a good grade of oil. Use 10W-30 or 10W-40. Do not use 50 weight racing oil! This is leftover thinking from the past. Oil has improved tremendously over the years and knowledge has improved as well—lightweight oil is the way to go. The oil pump cannot move heavy oil quickly, meaning that the engine will run for a few moments without full lubrication. You will definitely avoid problems if you stay away from heavy oil.

You may have heard, probably from old-timers, that it's best to use nondetergent oil for break-in. This is also archaic thinking. Nondetergent oil was used in the old days when rings were tough to break in. The oil would allow the residue from rings and so forth to act as an abrasive to help the rings seat. Keep in mind that this residue also went through the whole oil system and past the bearings. Abrasives and bearings do not mix. So, for break-in, just use the regular oil that you will run.

The oiling system must be primed with the distributor in place. You will get oil pressure without it, but the right lifter bank will go without lubrication. The distributor is part of the oil passage and must be there to complete the path. You can buy a priming tool or you can make one out of an old distributor.

To make a priming tool, place the old distributor on the workbench, supporting the gear end on a block of wood. With a hammer and punch, drive out the roll pin that holds on the gear. Remove the vacuum advance, rotor, breaker plate, and weights. Pull out the shaft; the plate with the pins for the weight springs will still be attached—this has to come off. To remove it you have to grind the swedged-over area in the center that holds on the plate. Once this is off, reinstall the gear with the roll pin. Using a bench grinder, grind the gear smooth.

Now, on the opposite end, grind the shaft so that it fits into a 3/8in drill. Clean the housing, then slide the shaft back into the

Use good quality oil and a good oil filter. Discount filters and off-brand oils are not recommended. Spend a few extra dollars now and save a lot of dollars down the road.

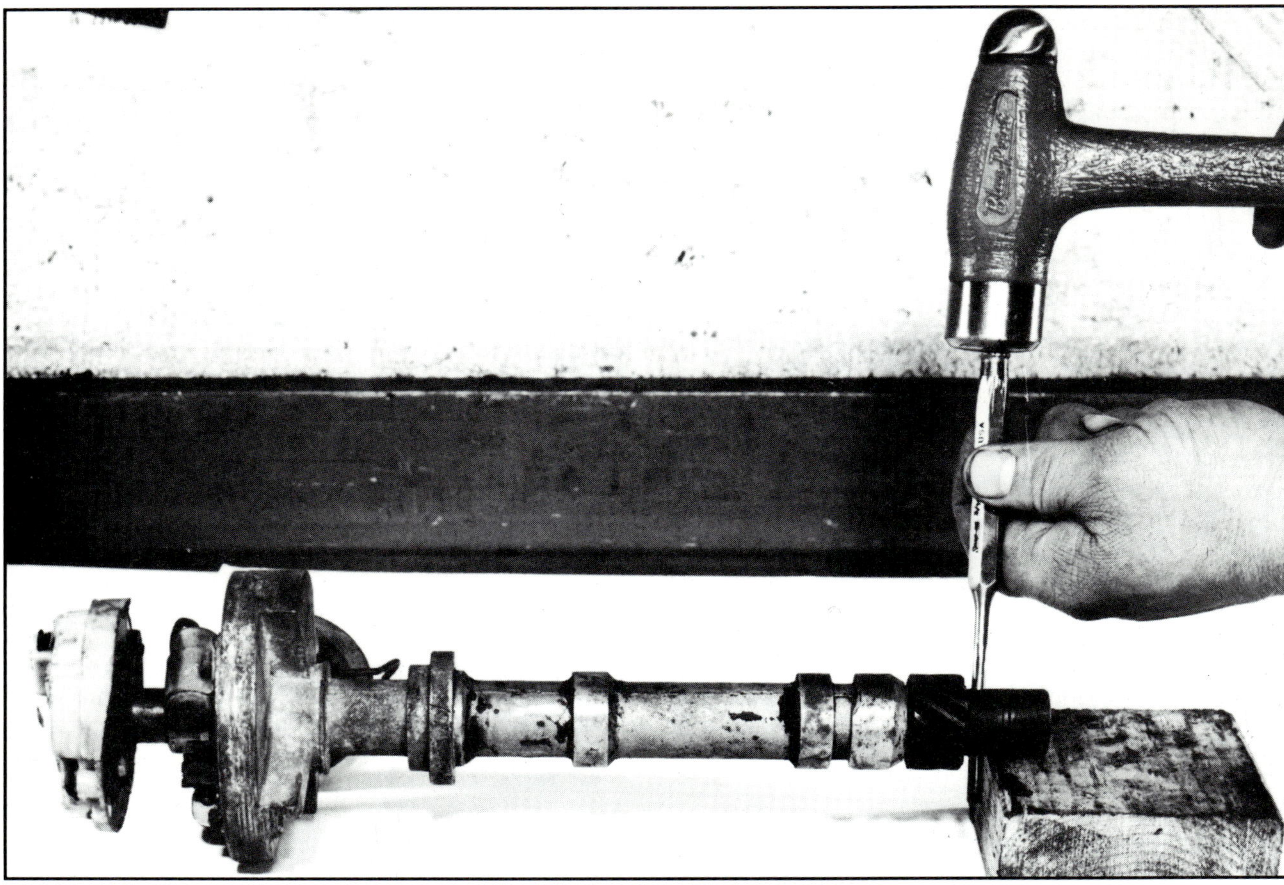

The first step in converting an old distributor to an engine oil primer is to remove the roll pin that holds on the gear. An appropriate size punch and a hammer will do.

now bare distributor body. Use a washer and a small hose clamp to retain the shaft in the housing. There you have it, a cheap, homemade engine oil primer. Slide it into place in the engine, attach an electric drill, hook up a mechanical oil pressure gauge, and you can prime the engine, resting assured that all passages will receive oil.

When you are priming, check the pressure gauge and make sure oil pressure is where it should be. With the drill wide open, we hit 60psi. If pressure is unusually low, find out why—right now—and fix it. After the engine is fired, it is too late.

Initial Fire-Up

A defective carburetor can ruin a new engine in a matter of minutes. Is the carburetor good? If it didn't work before you pulled the engine out of the car, it won't work now. If the power valve is damaged or the floats are sticking, the carb can flood the engine quickly. This will wash down the cylinder walls with gasoline, thus washing away the lubricating oil film. Ruined rings and scored cylinder walls are the result. Make sure that carb is good!

Have a timing light already hooked up, and set a close timing. To get the timing close, rotate the engine until it is at TDC (top dead center) on the power stroke—both valves will be closed. Set the distributor so that the rotor is pointing at the terminal for the number 1 plug wire. Turn the distributor counterclockwise 1/4in. (The engine should fire just before TDC.) This will be close enough to allow the engine to start.

One person will have to hold the timing light and one person will have to hit the switch and monitor the oil pressure. Fill the carburetor float bowl through the vent tube. Make sure the accelerator pump works. The engine should fire almost instantly. If it doesn't, there is something wrong. *Do not continue cranking.* Prolonged cranking will wreck a new camshaft and lifters. If the engine doesn't start in a few revolutions, back up and find out why.

As soon as the engine starts, bring the rpm up to 2000–2500. Idling during the first 15 minutes of running will destroy the cam and lifters.

Setting the Timing

After firing up the engine, the person with the timing light should immediately set the timing. On a race engine, full advance will be in at this rpm so set the timing at 38–40 degrees. On a street engine, you can drop the idle briefly to about 800rpm and set the timing at 4–6 degrees before top dead center (BTDC).

If you need to adjust the distributor to achieve correct timing, do not let the engine idle, but bring it back up to 2000rpm or so. Only when you are ready to make the timing check again should you let the engine idle down.

Troubleshooting Problems

Watch for oil and water leaks and monitor oil pressure. Also, inspect the fuel lines and fittings closely for leaks—a leak here could destroy your work and the whole car. Fire is a real possibility as well. You should have a fire extinguisher close at hand that can fight fuel fires. A heavy towel to smother carburetor flames is an excellent idea. After about 15 minutes of running, the engine can be idled down. Now you can set the idle speed where you want it and check the timing again.

Let the engine idle for a bit and run the rpm up and down a few times. The car is now ready to drive. Nowadays, there is no need for extended break-in time. If your engine is going to be driven on the street, avoid any full-out accelerations for a while, though. If it is going into a race car, it is ready to run.

At this point, let the engine cool down completely; this will take several hours, so it is best to let it sit overnight. If you are using a solid-lifter cam, reset the valve lash using the same procedure as described in engine assembly. This will have to be done several times.

Also, the head bolts should be retorqued. Gasket manufacturers claim that this is not necessary with the new gasket technology, but the fact remains that after initial heating and cooling, the head bolts often will not be on spec. It is especially important to retorque on a high-compression engine and/or one that will see hard duty. Follow the tightening sequence described in the assembly chapter.

Before the first race you may want to change spark plugs. The ones used for break-in may be slightly fouled—not enough to

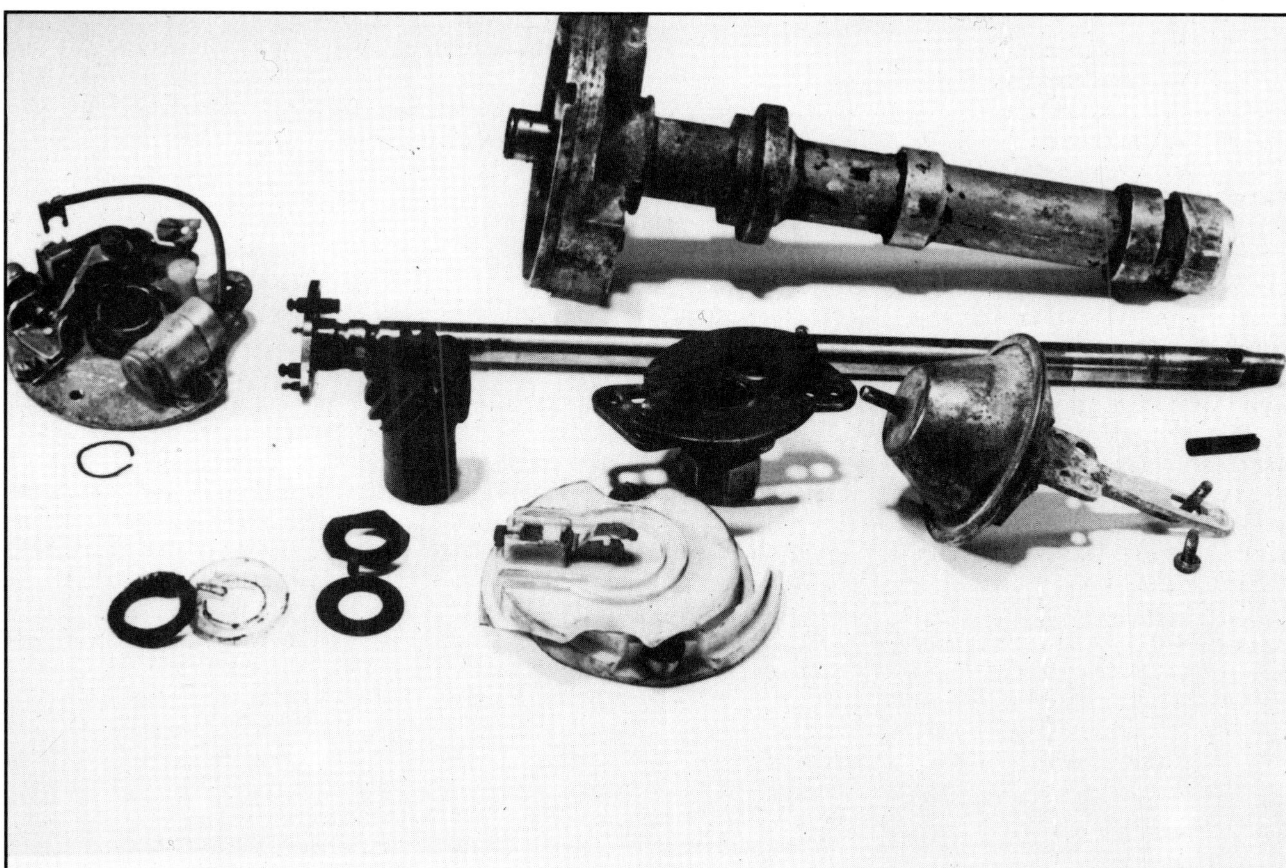

Here are all the pieces that must be removed.

prevent them from working, but they may not be up to full efficiency. The engine runs rich when first started and also uses a little oil in the first few minutes.

Change oil after the first 500 miles or after the first race. When you remove the filter, inspect it for bits of shiny metal. No need to cut open the filter, just look in on the top where all the little holes are. This is the area the oil reaches first after it leaves the block, so any particles will be caught here. If there is any contamination, find out what is wrong before the engine is destroyed.

If the car has open headers, internal engine sounds can be difficult if not impossible to hear. A long screwdriver is a cheap stethoscope. Put the handle end up to your ear and rest the other end on a valve cover or the intake manifold. You should hear a sewing machine-like sound. If you hear a knocking sound or any sharp blows coming from within, there is something wrong. Stop the engine and investigate.

Remember, you can make or break your new engine in a matter of minutes on initial startup. Do not crank it excessively; if it doesn't start within a few revolutions, then something is not right. Make sure it has fuel, the timing is set close, and it is getting spark. Do not idle the engine for the first fifteen minutes. Check the timing immediately upon firing, monitor the oil pressure, check closely for fuel leaks. If you still can't locate the problem, ask for help from a *knowledgeable* person. Do these things and all the hard work and money you invested in your new engine will pay rich rewards.

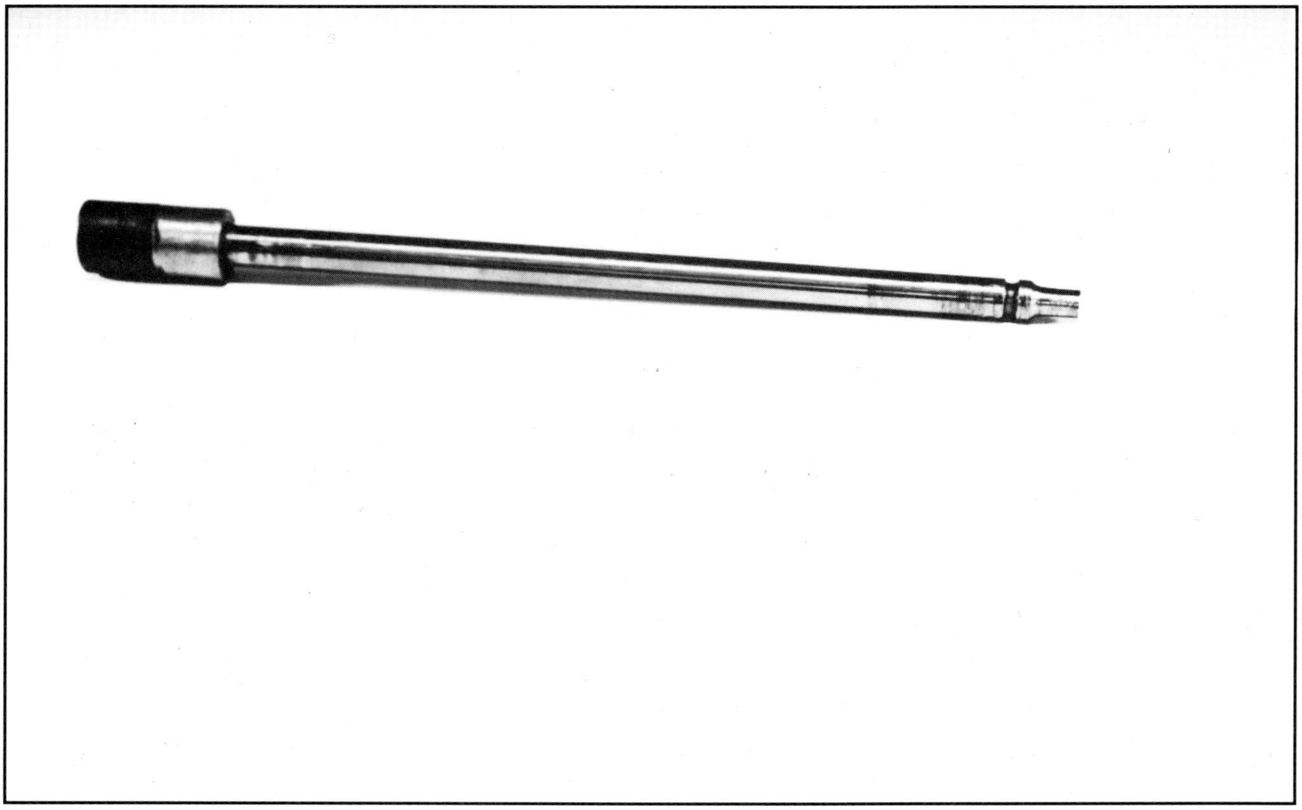

Reinstall the gear and then grind off the teeth. Also, grind down the opposite end of the shaft so that it will fit your 3/8in drill.

Slide the shaft back into the housing and retain it with a washer and a hose clamp. Your homemade engine primer is finished.

Hook up an oil pressure gauge, install the primer, and spin it with a drill. If everything is in order, the engine will reach oil pressure. Here we have about 50lb. When the drill was turned wide open, pressure hit 60lb.

Engine Accessories 9

The engine itself is only part of the package that makes the car go. You can improve performance and safety by adding on the right parts to enhance the engine.

Carburetors

The carburetor you use will make a huge difference in power. If you can afford a professionally prepared carburetor, tailored to your specific application, it will be worth horsepower. At the least, you should use a good Holley carb, one that is new or in good condition. An improperly functioning carburetor, as well as costing power, may actually damage the engine. As stated previously, a carburetor that floods the cylinders with gasoline will quickly destroy engine parts.

If you are racing your car and the rules allow it, run alcohol. Alcohol is good for about 50 additional horsepower and it runs cooler than gasoline, therefore helping to control overheating problems. It also does not detonate like lower octane gasoline.

A negative aspect of alcohol is initial cost. A special carburetor has to be used, one that is calibrated for the fuel—it takes about twice as much alcohol as gasoline to run the engine. These carburetors are expensive. The fuel cell must be compatible with alcohol, as must the rest of the components in the system.

Alcohol has no lubricating properties. Therefore, cylinder heads must have hard exhaust seats and guides. The Dart heads come with hard seats, but if you are using a set of old Chevy double-hump castings, hard seats will need to be installed. Bronze guides also must be used with alcohol. Cast guides, even hardened ones, will wear out. Cast-iron Dart heads do not have bronze guides, so have them installed at the machine shop. As explained in chapter 4, the best bronze guides are the 0.500in replacement types and are *solid* bronze, not cast iron with a bronze liner.

Whether or not you choose to run alcohol will depend upon your pocketbook. There are definite benefits, but the initial cash outlay is considerable.

Exhaust

Headers are thought of as being synonymous with horsepower. For the racetrack they are a definite advantage. For the street, they are a definite headache. They blow gaskets, leak, make it hard to get to the spark plugs and starter, and they rust out. The stock manifolds do not present any of these problems, however, and a stock-appearing exhaust system can be retained.

Few things can make an otherwise nice-looking car look shabbier than mufflers about to drag on the ground. Install flow-through-type mufflers, the quiet ones, so that you do not draw the wrong kind of attention. If you want your stock cast-iron exhaust manifolds to flow better and don't mind spending the money, they can be worked over by specialists such as Brzezinski Racing Products, Inc. You will have header-like performance with the convenience of a stock exhaust system.

For racing, headers are worthwhile. The engine is easier to get at in a race car, there is no exhaust system to worry about, the car presumably is not driven in wet weather to rust the headers, and a small exhaust leak won't be heard.

Racing headers do allow the engine to make power. As a general rule, small-tube headers help bottom-end torque and big tubes help mid- and upper-end power. Also, longer collectors generally help low- and mid-range torque while shorter ones help upper-end. Longer collectors also make it easier to control the engine with the throttle.

You can actually eliminate about 6lb of weight off the front of the car by cutting out the metal flange between the header tubes. Save weight by eliminating anything that is not needed, but at the same time, be safety conscious. Do not cut corners on safety equipment—any weight savings here are not worth the risk to your life or the lives of your competitors.

Ignition

Invest in a good set of plug wires. Inexpensive wires go bad in a short time, making the engine run terribly. It will be down on power and will pop and miss.

A good ignition system that maintains accurate timing and produces a good, hot spark is vital for producing horsepower. An MSD ignition system is an excellent choice.

There is one problem with a hot ignition setup like this, though. The power is so great that it tries hard to follow the path of least resistance. On a small-block Chevrolet, cylinders 1 and 2 and cylinders 5 and 6 tend to crossfire. On a car with headers, if crossfiring is a problem, the pipes on the cylinders that are "laying down" will be cool in comparison to the others. This causes pre-ignition, meaning that the cylinder is firing too much ahead of time. The results of pre-ignition in a powerful engine working hard can be broken main caps and a broken crankshaft, which lead to ultimate destruction.

The cure for this problem is simple. Install a "fat cap" on the distributor. This is an overside cap that spreads out the wires so that crossfiring is eliminated.

Belts and Pulleys

For racing, to keep the belts from coming off the engine invest in some deep-groove racing pulleys. When you back off the throttle, the belts will flop and go sailing off with regular pulleys. The engine won't last long, even on the racetrack, without the water pump pulley.

Engine Cooling

The radiator is a vital piece of equipment—a good radiator will keep the engine running and keep it running well. On the other hand, a radiator that is plugged up will allow the engine to overheat, possibly to the point of severely damaging it from detonation or even actually melting it.

A stock radiator is fine and will do the job well. While the engine is out of the car, take the radiator to a shop and have them inspect and clean it. Then you can rest assured that it will do the job of keeping your new engine cool.

Also, always install a new thermostat; never put a used thermostat on a new engine. That little $4 part can ruin your $1,000 engine. If your car is not computer equipped, a 180 degree Fahrenheit thermostat is right for the street. A computer-equipped vehicle may not perform correctly with a 180 degree thermostat, so check your owners manual or with your car dealer. Also, you may want to check your state emissions laws regarding thermostats and any other modifications. For the racetrack, do not use a thermostat, but experiment with different restrictors to achieve correct temperature.

A stock fan also will do the job. Be sure to retain the stock shroud along with it. If you decide to use a flex fan, check for clearance between fan and radiator, fan and shroud, and fan and any other object close by. Those flex fans really do flex, in all directions. More than one has torn up a radiator when the engine was revved.

Transmission and Driveline Safety

If the engine will be turned over 5000rpm, a safety shield must be used over the clutch and flywheel or, in the case of an automatic transmission, over the torque converter. These things can and do fly apart sometimes. If one explodes, it will come through the floorboard and remove your toes. Pieces also may fly out the side of the car or bounce off the ground and injure or kill a spectator. Serious business here.

For a stick-shift situation, buy an approved steel bellhousing designed to contain the clutch and flywheel in the event that they explode. For an automatic transmission, a ballistics blanket should be used. It goes around the transmission and absorbs pieces of torque converter in an explosion. Most racetracks and sanctioning bodies wisely require these safety items.

If you run an automatic transmission, invest in a shift kit. It will improve the performance of the transmission and help it transfer the engine's power to the rear wheels.

In either an automatic or stick-shift, check into using a synthetic lubricant. If the transmission manufacturer suggests it, try it. The synthetic fluids and oils do not break down with heat as do petroleum-based lubricants. Be sure to check with manufacturer's specs before making the change, though.

Clutch

If rotating mass is reduced, the car will accelerate quicker because less power will be used to get the mass rotating. One area where you can reduce rotating mass is in the clutch.

The smaller the diameter, the less the mass. Double- and triple-disc clutches have smaller diameters, yet have as much and oftentimes more friction area for gripping.

In a stock car, in a nonshifting application, a coupler can be used in place of a clutch. Ram Automotive makes a coupler that is only 4 3/4in in diameter and weighs just 5lb, thus drastically reducing rotating mass. The coupler is controlled by a clutch pedal and may be disengaged for starting and stopping the car, thus retaining the convenience of a clutch.

Driveshaft Parts

Install heavy-duty U-joints in the driveshaft. The heavy-duty joints do not have grease fittings, thus making them stronger. A broken U-joint can be costly in damage to the rear end and maybe even to the floorboard and driver. If you race the car or plan on some hard passes out on the blacktop road, invest in driveshaft safety loops. They are designed to keep the shaft from coming through the floorboard or, in case of front joint failure, digging into the ground and vaulting the car over.

Rear End Ratios

Installing lower rear-end gears will help performance tremendously. For street driving, lower gears will hurt fuel mileage, so some kind of compromise is needed. The 3.08:1 to 3.73:1 gears are streetable and will help the acceleration performance. These gears do not sound low when compared to 5.14:1 racing gears, but you have to compromise for drivability and fuel mileage on the street.

The lower the gears, the more rpm the engine will turn at a given speed. The faster the engine turns, the quicker it will wear out, too. All these factors have to be weighed when selecting a gear. In the end, it depends upon what is most important to you.

As with the transmission, use synthetic oil in the rear end. Again, check with the manufacturer's specs to make sure it is acceptable.

Tires and Rims

If you can go with a wider rim for the tires, you can gain a little footprint area. The wider rim pulls out the sidewalls, making the tread area fatter. This puts more of the tire on the ground for more traction. A tire with a softer compound will grip the pavement better, but also will wear out more quickly. As with the rear-end gears, it is a situation of compromises and choices.

Although they cost more than a set of stock steel rims, lightweight racing wheels are a benefit. This is because unsprung weight, that is, anything not supported by the springs, such as wheels, should be kept to a minimum. If you do buy a set of aftermarket wheels, make sure they really are lighter than stock rims. Not all of them are. Also, make sure that what you buy is safe for street use and will withstand cornering forces. The rims have to be approved by the Department of Transportation (DOT) to be safe on the street.

Refer to good chassis books and ask advice of chassis experts for setting up your car for peak performance. Listen to what they have to say and follow through. The engine, no matter how powerful, is only *part* of the overall package for performance. Treat your car as a collection of components working together, all of them of equal importance, to get the most performance possible.

Results, Conclusions, and Other Useful Information 10

355 Results

Our test engines performed to perfection on the dynamometer. The little 355, when idled down, purred with only the slightest lope from the cam. It is a smooth-running engine, yet there is the distinct sound of lurking horsepower waiting to be unleashed when you hit the throttle.

The 355, as stated before, is an everyday driver, one that will go get the groceries, take you on a trip across the country, yet has the power and torque in the right rpm ranges to make it a really fun engine to drive. The response is there when you hit it.

Also, as explained earlier, you can duplicate this 355 (except for the 0.030in overbore) by installing on your car's engine the World Products Dart 305 heads, the Elgin 903-P camshaft, and the Weiand dual-plane manifold. Add a good Holley carburetor and the results should be similar to what we achieved here.

The dyno figures presented here are not inflated; they are the real, corrected figures. The dyno operator can feed the parameters into the computer to make it read whatever he or she wants, if the goal is inflated figures to create an impression. There is no such thing as a 500hp small-block put together from stock and nearly stock parts, like we have here. It takes real racing components to attain these figures. Therefore, our dyno results may seem mild compared to some "tests" you have read about elsewhere.

Dyno readings have to be corrected according to air temperature, air pressure, and humidity

Our little 355 on the dyno, all hooked up and ready to come to life. It fired right up and after a little break in time, hit 314hp and 327lb-ft of torque. All this out of a streetable, smooth-running engine.

to give figures which can be compared to readings from another engine tested somewhere else.

355 Dyno Results

RPM	Power	Torque
3500	218hp	327lb-ft
4000	249	327
4500	275	321
5000	281	294
5500	314	280
6000	256	257

The 355ci engine puts out more than 100hp over a stock four-barrel 350. Larry Wright, our dyno operator, has tested many Chevy engines in his classroom and they all fall in the range of 190 to 210hp. When you consider the mild and inexpensive parts that went into the engine, this is a considerable increase.

We used a stock 600cfm Holley carburetor and pump gasoline. The headers are not special dyno headers, but a set from a Camaro. Therefore, the headers have some rather sharp bends in them for clearing chassis components. A set of dyno headers with long, straight, equal-length tubes should produce more power but you don't run these in a car anyway. Power figures will vary up or down depending upon the carburetor and exhaust system you use.

Another important factor is the relationship of the horsepower and torque. The horsepower peaks at the shift point, and the torque peaks at about the rpm the engine drops to when you shift. When you make the gear change the torque will pull you to get you going, then as the torque falls off, the horsepower comes into play. This is the perfect setup for a good-running street engine. It would be great fun to drive it!

406 Results

For all practical purposes, the 406 is a race-only engine. The vacuum produced from the racing cam is too low for operating accessories on a street car. Also, the off-idle response would be poor, making it a hassle to drive in traffic. On the racetrack where rpm is maintained, however, this engine will be a strong performer.

Just as with the 355, the torque increases as rpm drops—this will help pull the car off the corners and get it going until the horsepower takes over. This engine will work great in a Saturday night short-track car; it

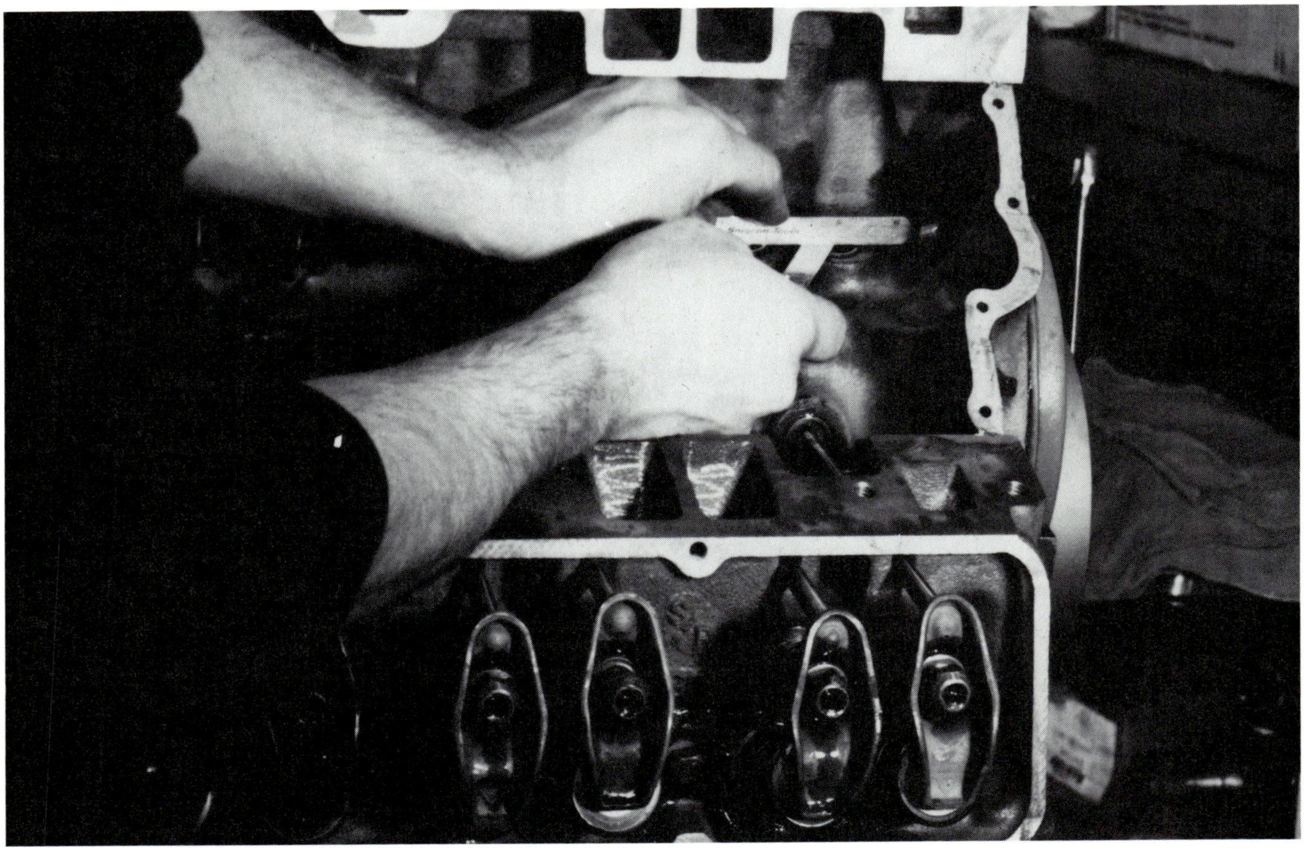

Checking cam timing with the feeler gauge method. Though not quite as accurate, this method does not require expensive tools to do the job. It will give you a very close estimate of where your cam is in relation to the crankshaft.

would do its job in a street stock, a modified, or a pro-stock. It would also be at home between the rails of a bracket racer producing consistent, dependable performance. Granted, it won't keep up with a $10,000 race engine, but for the investment it is a winner.

406 Dyno Results

RPM	Power	Torque
4000	310hp	407lb-ft
4500	340	395
5000	361	380
5500	381	363
6000	372	325
6500	328	265

For the dyno test, an out-of-the-box bone-stock 750cfm Holley carburetor was used with a 1in spacer underneath. The same Camaro headers that were on the 355 were used on this engine as well. A professionally prepared and flowed carburetor should easily put this engine over 400hp. Alcohol fuel would push the figure even higher, approaching the 450hp mark. Even with the basics in induction and headers, as we used here, this engine is a strong runner.

You can build either one of these engines, or one very similar, if you take your time and follow our suggestions. You may have noticed an absence of technical information in this book. Technical specs were left out intentionally, since this book is meant to be a practical, hands-on guide for the everyday backyard enthusiast working on a budget with stock parts. An overabundance of technical and engineering information can become confusing, though it does have its place. We simply felt it was not appropriate here. We want you to be able to take this book out in the garage and use it in *your* engine building.

Sources

Brzezinski Racing Products, Inc.
N50 W23001 Betker Drive
Pewaukee, WI 53072

Competition Cams
3406 Democrat Road
Memphis, TN 38118

Competition Products
3090 Oregon Drive
Oshkosh, WI 54901

Elgin Industries
1100 Jansen Fram Drive
Elgin, IL 60123

E-Loy
S. B. International, Inc.
2108 Utopia Avenue
Nashville, TN 37211

Holley Carburetor
11955 E. Nine Mile Road
Warren, MI 48090

Iskenderian Racing Products
16020 S. Broadway
Gardena, CA 90247-9990

JR Manufacturing, Inc.
6200 Hwy 12 East
Eau Claire, WI 54701

MSD Ignition
Autotronic Controls Corporation
1490 Henry Brennan Drive
El Paso, TX 79936

RAM Automotive
201 Business Park Boulevard
Columbia, SC 29203

Sealed Power
100 Terrace Plaza
Muskegon, MI 49443

Silv-O-Lite Pistons
United Machine and Tool Co.
4909 Goni Road
Carson City, NV 89706

Weiand Automotive
P.O. Box 65301
Los Angeles, CA 90065

World Products, Inc.
80 Trade Zone Court
Ronkonkoma, NY 117795

Index